When Less is More

Visualizing Basic Inequalities

© 2009 by
The Mathematical Association of America (Incorporated)
Library of Congress Catalog Card Number 2008942145
ISBN 978-0-88385-342-9
Printed in the United States of America
Current Printing (last digit):
10 9 8 7 6 5 4 3 2 1

The Dolciani Mathematical Expositions

NUMBER THIRTY-SIX

When Less is More

Visualizing Basic Inequalities

Claudi Alsina
Universitat Politècnica de Catalunya

Roger B. Nelsen
Lewis & Clark College

Published and Distributed by
The Mathematical Association of America

The DOLCIANI MATHEMATICAL EXPOSITIONS series of the Mathematical Association of America was established through a generous gift to the Association from Mary P. Dolciani, Professor of Mathematics at Hunter College of the City University of New York. In making the gift, Professor Dolciani, herself an exceptionally talented and successful expositor of mathematics, had the purpose of furthering the ideal of excellence in mathematical exposition.

The Association, for its part, was delighted to accept the gracious gesture initiating the revolving fund for this series from one who has served the Association with distinction, both as a member of the Committee on Publications and as a member of the Board of Governors. It was with genuine pleasure that the Board chose to name the series in her honor.

The books in the series are selected for their lucid expository style and stimulating mathematical content. Typically, they contain an ample supply of exercises, many with accompanying solutions. They are intended to be sufficiently elementary for the undergraduate and even the mathematically inclined high-school student to understand and enjoy, but also to be interesting and sometimes challenging to the more advanced mathematician.

MAA Service Center
P.O. Box 91112
Washington, DC 20090-1112
1-800-331-1MAA   FAX: 1-301-206-9789

We learned from them the first inequality in life:
Love is more important than anything else.

Dedicated to our beloved mothers

María and Ann

in memoriam.

Contents

Preface

The worst form of inequality is to try to make unequal things equal.

Aristotle

The fundamental results of mathematics are often inequalities *rather than* equalities.

Edwin Beckenbach and Richard Bellman

Inequalities permeate mathematics, from the *Elements* of Euclid to operations research and financial mathematics. Yet too often, especially in secondary and collegiate mathematics, the emphasis is on things equal to one another rather than unequal. While equalities and identities are without doubt important, they do not possess the richness and variety that one finds with inequalities.

The objective of this book is to illustrate how use of visualization can be a powerful tool for better understanding some basic mathematical inequalities. Drawing pictures is a well-known method for problem solving, and we would like to convince you that the same is true when working with inequalities (recall George Pólya's advice "Draw a figure ... "). We will show how to produce figures in a systematic way for the illustration of inequalities; and open new avenues to creative ways of thinking and teaching. In addition, a geometric argument can not only show two things unequal, but also help the observer see just how unequal they are.

Visual arguments known as *proofs without words* are published regularly in *Mathematics Magazine*, *The College Mathematics Journal*, and other publications. Some involve inequalities, and appear in [Nelsen, 1993 and 2000]. We have also published a book [Alsina and Nelsen, 2006] on how to create images to help one understand mathematical ideas, proofs, and arguments. *When Less is More* follows the tradition of these resources, but focuses on basic mathematical inequalities, primarily, but not exclusively, from geometry.

Since our main objective is to present a methodology for producing mathematical visualization of inequalities, we have organized the book as follows. Following a short introduction discussing the importance of inequalities in mathematics, we present, in nine chapters, methods for creating pictures to illustrate (or even prove) inequalities, followed by examples and applications of these methods. Each chapter ends with a collection of challenges that the reader may work in order to better understand the methodology of the chapter, and to apply that methodology to additional inequalities.

Next we present solutions to all of the challenges in the chapters. *When Less is More* concludes with a list of notation and symbols, references, and a complete index. We invite readers to become actively involved in the process of visualizing inequalities and to find new or more elegant proofs than those presented in this book, and to apply the methods we describe to prove other inequalities, classical and new.

As noted earlier, inequalities appear in nearly every branch of mathematics, both pure and applied. The fact that we concentrate on geometric inequalities is partially motivated by our hope that secondary and collegiate teachers might use our pictures with their students. Teachers may wish to use one of the drawings when an inequality arises in the course. Alternatively, *When Less is More* might serve as a guide for devoting some time to inequalities and problem solving techniques, or even as part of a course on inequalities.

Special thanks to Rosa Navarro for her superb work in the preparation of the final draft of this manuscript. Thanks too to Underwood Dudley and the members of the editorial board of the Dolciani Mathematical Expositions for their careful reading of an earlier draft of the book and for their many helpful suggestions. We would also like to thank Elaine Pedreira, Beverly Ruedi, and Don Albers of the MAA's book publication staff for their expertise in preparing this book for publication. Finally, special thanks to friends in Argentina, Spain, and the United States for their willingness to explore these ideas and techniques with us, and for encouraging us to work on this publication.

Claudi Alsina
Universitat Politècnica de Catalunya
Barcelona, Spain

Roger B. Nelsen
Lewis & Clark College
Portland, Oregon

Introduction

"When I use a word," Humpty Dumpty said in rather a scornful tone, "it means just what I choose it to mean— neither more nor less."
Lewis Carroll, *Through the Looking-Glass* (1871)

Why study inequalities? Richard Bellman [Bellman, 1978] answered:

There are three reasons for the study of inequalities: practical, theoretical, and aesthetic.

In many practical investigations, it is necessary to bound one quantity by another. The classical inequalities are very useful for this purpose.

From the theoretical point of view, very simple questions give rise to entire theories. For example, we may ask when the nonnegativity of one quantity implies that of another. This simple question leads to the theory of positive operators and the theory of differential inequalities. The theory of quasilinearization is a blend of the theory of dynamic programming and that of positive operators. This is typical of mathematics. Each new theory uses parts of existing theories.

Another question which gives rise to much interesting research is that of finding equalities associated with inequalities. We use the principle that every inequality should come from an equality which makes the inequality obvious.

Along these lines, we may also look for representations which make inequalities obvious. Often, these representations are maxima or minima of certain quantities.

Again, we know that many inequalities are associated with geometric properties. Hence, we can go in either direction. We can find the geometric equivalent of an analytic result, or the analytic consequence of a geometric fact such as convexity or duality.

Finally, let us turn to the aesthetic aspects. As has been pointed out, beauty is in the eye of the beholder. However, it is generally agreed that certain pieces of music, art, or mathematics are beautiful. There is an elegance to inequalities that makes them very attractive.

From the historical point of view, since inequalities are associated with order, they arose as soon as people started using numbers, making measurements, and later, finding approximations and bounds. Thus inequalities have a long and distinguished role in the evolution of mathematics.

Inequalities as a field of study

> *It was a wise man who said there is
> no greater inequality than the equal
> treatment of unequals.*
>
> Felix Frankfurter

While it is obvious that inequalities are present in nearly every branch of mathematics, the study of inequalities has become a field in itself. The celebrated book *Inequalities* by G. H. Hardy, J. E. Littlewood, and George Pólya, first published in 1934, is a classic in the field.

G. H. Hardy   J. E. Littlewood   G. Pólya

Later, popular monographs such as *An Introduction to Inequalities* by Edward Beckenbach and Richard Bellman and *Geometric Inequalities* by Nicholas D. Kazarinoff (now published by the Mathematical Association of America) broadened the appeal of the subject. The book *Geometric Inequalities*, written by O. Bottema and his colleagues in 1968 with over 400 inequalities, became a valuable reference, as has its successors *Recent Advances in Geometric Inequalities* by D. S. Mitrinovic et al., published in 1989 with several thousand inequalities, and P. S. Bullen's 1998 book *A Dictionary of Inequalities*. Lastly, we mention *Inequalities: Theory of Majorization and its Applications* by A. W. Marshall and I. Olkin, with its contribution of new approaches and applications.

As in other branches of mathematics, a number of symposia on "general inequalities" made possible periodic meetings of mathematicians interested in inequalities, and produced published proceedings and many journal papers devoted to research on inequalities. As of 2009, an Internet search on the word "inequalities" will yield over 12 million web pages.

Inequalities in the classroom

> *Inequality is the only bearable thing,*
> *the monotony of equality can only*
> *lead us to boredom.*
>
> Francis Picabia

In 1611, the painter and architect Lodovico Cardi (1559–1613), also known as Cigoli, wrote to Galileo Galilei (1564–1642) (see, e.g., [De Santillana, 1955]):

A mathematician, as good as he may be, without the support of a good drawing, is nothing but a half-mathematician, but also a man without eyes.

In teaching mathematics, visualization is essential to develop intuition and to clarify concepts. The need for inequalities in the classroom arises as soon as the ordering of numbers is considered. While the ordering of the natural numbers $0, 1, 2, \ldots$, offers no problems, as soon as integers are introduced we need to consider how multiplication by a negative integer changes order. With fractions, order becomes more sophisticated, since the concept of "the next one" does not apply. In addition, there are several meanings and representations for rational numbers (ratios, proportions, slopes, etc.). In the case of irrational numbers, approximations by sequences of rational numbers (e.g., decimal expansions) necessitate the consideration of inequalities. As Henri Poincaré wrote [Poincaré, 1952]: "if arithmetic had remained free from all intermixture with geometry, it would never have known anything but the whole number. It was in order to adapt itself to the requirements of geometry that it discovered something else."

Clearly it is important to master the various inequalities for real numbers since, beyond arithmetical considerations, they play a major role in geometrical and functional inequalities of all types. In plane and solid geometry, inequalities appear naturally when comparing measures (lengths, areas, volumes, ...), in determining the existence or nonexistence of particular figures, and solving optimization problems in a variety of settings. In addition, many classical theorems in geometry may be stated and proved by means of inequalities. The focus of this book is the approach to inequalities by means of geometric drawings.

In the case of algebraic inequalities one may proceed to give proofs via algebraic manipulations, but as we shall see in the following chapters, many of the basic algebraic inequalities also admit a visual proof by representing the algebraic quantities involved in a suitable fashion.

Thus the methods encountered in the book have a pedagogical value: visual representations are excellent tools for developing intuition, for better understanding many concepts and results, and for enjoying proofs.

Symbols for inequalities

The symbol $=$ for equality appears to have been introduced by Robert Record (c. 1510–1558) in his book *The Whetstone of Witte*, published in 1557. Record used an elongated form, and wrote "I will sette as I doe often in woorke use, a paire of parralles, or Gemowe lines of one length, thus $=$, bicause noe 2 thynges, can be moare equalle." This symbol did not appear in print again until 1618, but soon thereafter replaced words commonly used to express equality, such as *aequales* (often abbreviated *aeq*), *esgale*, *faciunt*, *ghelijck*, and *gleich*.

The symbols $>$ and $<$ to denote strict inequality appeared a few years later, in *The Analytical Arts* by Thomas Harriot (1560–1621), published in 1631. Harriot states the meaning for $>$ and $<$ quite clearly: *Signum majoritatus ut a $>$ b significet a majorem quam b*, and *Signum minoritatus ut a $<$ b significet a minorem quam b* ($a > b$ means a is larger than b, and $a < b$ means a is smaller than b). According to Eves (1969), during a visit to the colony of Virginia in 1685 Harriot saw a native American with the design $\gtrless$ on his left shoulder blade, and speculates that Harriot developed his inequality symbols from this design. Eves also suggests this may not be true.

While these data about Harriot constitute the standard historical record, there has been some discussion about Harriot's invention since *The Analytical Arts* may have been the work of others. Nevertheless, 1631 is the birth date for $>$ and $<$. Pierre Bouguer (1698–1758) used $\geqq$ and $\leqq$ in 1734, while John Wallis (1616–1703) used similar notation but with the bars above the inequality symbols. Symbols crossed with a vertical or inclined bar to denote "not equal to," "not greater than," and "not less than" were employed by Euler.

Basic properties of inequalities for real numbers

- *Reflexivity*: $a \leq a$

- *Antisymmetry*: $a \leq b$ and $b \leq a$ imply $a = b$

- *Transitivity*: $a \leq b$ and $b \leq c$ imply $a \leq c$

- *Trichotomy*: Either $a < b$ or $a = b$ or $a > b$

- *Addition*: $a \leq b$ implies $a + c \leq b + c$

- *Multiplication*: $a \leq b$ and $c \geq 0$ implies $ac \leq bc$
  $a \leq b$ and $c \leq 0$ implies $ac \geq bc$

There is always inequality in life.
John F. Kennedy

CHAPTER **1**

Representing positive numbers as lengths of segments

A simple, common, but very powerful tool for illustrating inequalities among positive numbers is to represent such numbers by means of line segments whose lengths are the given positive numbers. In this chapter, and in the ones to follow, we illustrate inequalities by comparing the lengths of the segments using one or more of the following methods:

1. *The inclusion principle.* Show that one segment is a subset of another. We will generalize this method in the next chapter when we represent numbers by areas and volumes and illustrate inequalities via the subset relation.

2. *The geodesic principle.* Use the well-known fact that the shortest path joining two points is the linear segment connecting them.

3. *The Pythagorean comparison.* Proposition I.19 in *The Elements* of Euclid states "In any triangle the side opposite the greater angle is greater." Thus in a right triangle, the hypotenuse is always the longest side. So to compare two segments, show that one is a leg and the other the hypotenuse of a right triangle.

4. *The triangle (and polygon) inequality.* Proposition I.20 in *The Elements* states "In any triangle the sum of any two sides is greater than the re-maining one." Thus when three line segments form a triangle, the length of any one of them is less than or equal to the sum of the other two (and similarly for polygons). This is a special case of the geodesic principle.

5. *Comparing graphs of functions.* If the graph of $y = f(x)$ lies above the graph of $y = g(x)$ over some interval of x-values, then for each

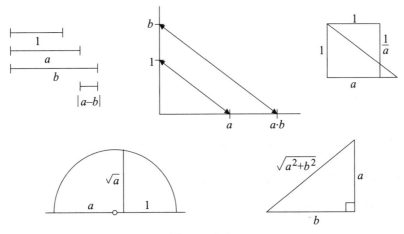

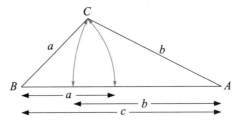

Figure 1.1.

x in the interval, the line segment joining $(x, f(x))$ to $(x, g(x))$ has nonnegative length, which establishes $f(x) \geq g(x)$.

In Figure 1.1 above we see how we can associate certain arithmetic operations on numbers with segments of given positive lengths.

1.1   Inequalities associated with triangles

If three positive numbers a, b, and c are the lengths of the sides of a triangle ABC, then $a + b > c$, $b + c > a$, and $c + a > b$ from the geodesic principle. The converse is also true from the inclusion principle. Without loss of generality we can assume $a \leq b \leq c$. Then only the first inequality $a + b > c$ is not trivial and is illustrated below, where we see that the side of length c is included within the union of two segments of lengths a and b. Indeed, this is the procedure one uses to construct a triangle with given side lengths using a straightedge and compass.

Figure 1.2.

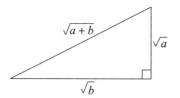

Figure 1.3.

This simple result has a number of useful consequences, especially when the triangle is a right triangle. For example, for positive a and b, we can take as the lengths of the sides of a right triangle the numbers $\sqrt{a}$, $\sqrt{b}$, and $\sqrt{a+b}$ to conclude that $\sqrt{a+b} < \sqrt{a} + \sqrt{b}$. See Figure 1.3. (If we allow a or b to be 0, then $\sqrt{a+b} \leq \sqrt{a} + \sqrt{b}$.) Functions f such as the square root that satisfy $f(a+b) \leq f(a) + f(b)$ are called *subadditive*, which we will have more to say about in Chapter 8.

In this chapter we will encounter various ways of averaging numbers, or of finding means. Perhaps the best-known mean is the *arithmetic mean*, which for two numbers a and b is $(a+b)/2$. Another mean is the *root mean square* or *quadratic mean*, defined as the square root of the arithmetic mean of the squares, which for two numbers a and b is $\sqrt{(a^2+b^2)/2}$. The root mean square is common in physics and electrical engineering where it is used to measure magnitude when quantities are both positive and negative, such as waves. In Figure 1.4 we use the triangle inequality twice to show that for positive numbers a and b, the root mean square lies between the arithmetic mean and $\sqrt{2}$ times the arithmetic mean [Ferréol, 2006], i.e.,

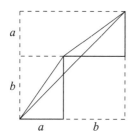

$$\sqrt{2}(a+b) \leq 2\sqrt{a^2+b^2} \leq 2(a+b)$$

$$\frac{a+b}{2} \leq \sqrt{\frac{a^2+b^2}{2}} \leq \frac{a+b}{\sqrt{2}}$$

Figure 1.4.

The same idea can be applied to three positive numbers a, b, and c to establish

$$\sqrt{2}(a+b+c) \leq \sqrt{a^2+b^2} + \sqrt{b^2+c^2} + \sqrt{c^2+a^2} \leq 2(a+b+c),$$

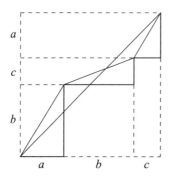

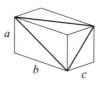

Figure 1.5.

as seen on the left in Figure 1.5 [Ferréol, 2006]. These inequalities give bounds for the sum of the side lengths of a triangle formed from the three diagonals of the faces of a rectangular box or parallelepiped, with edge lengths a, b, and c, as seen at the right in Figure 1.5.

1.2   Polygonal paths

The application of the geodesic principle to polygonal paths, as illustrated in Figures 1.4 and 1.5, can be extended to illustrate other inequalities. For example, for any positive numbers x, y, u, and v, we have

$$\sqrt{(x+y)^2 + (u+v)^2} \le \sqrt{x^2 + u^2} + \sqrt{y^2 + v^2},$$

as illustrated in Figure 1.6 [Kazarinoff, 1961]:

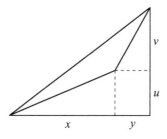

Figure 1.6.

This can be readily extended to n variables to obtain the following special case of *Minkowski's inequality* (Hermann Minkowski, 1864–1909) for positive numbers a_i and b_i [Shklarsky et al, 1962]:

$$\sqrt{\left(\sum_{i=1}^{n} a_i\right)^2 + \left(\sum_{i=1}^{n} b_i\right)^2} \le \sum_{i=1}^{n} \sqrt{a_i^2 + b_i^2}.$$

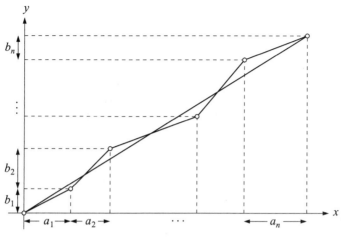

Figure 1.7.

1.3 *n*-gons inside *m*-gons

An *n-gon* is a polygon with *n* sides, i.e., a polygonal path, as those described above where the initial and terminal (or *extreme*) points coincide. If we draw one polygon inside another, there is an obvious area inequality, but is there also a perimeter inequality? In general the answer is no, since it is always possible to draw inside one polygon another one with an arbitrarily large perimeter.

But if we restrict ourselves to *convex* polygons, the answer is yes. Recall that an *n*-gon is convex if the line segment joining any two points on or in the *n*-gon lies entirely within the *n*-gon. Now consider a convex *m*-gon that contains a convex *n*-gon, as illustrated in Figure 1.8.

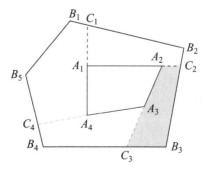

Figure 1.8.

Let $A_1, A_2, \ldots, A_n$ denote the vertices of the n-gon in clockwise order (letting $A_0 = A_n$ will be convenient), and similarly let $B_1, B_2, \ldots, B_m$ denote the vertices of the m-gon, with $B_0 = B_m$. Now, for each i from 1 to n, extend side $A_{i-1}A_i$ from vertex A_i until it meets the outer m-gon at a point C_i, and apply the polygon inequality to the polygonal region with vertices A_{i-1}, A_i, C_i, the vertices of the m-gon between C_i and C_{i-1}, and A_{i-1}. Using the absolute value symbol to denote length, we have

$$|A_{i-1}A_i| + |C_i A_i| \le |C_i B_1'| + \cdots + |C_{i-1}A_{i-1}|,$$

where the ellipsis on the right includes all the vertices B_j' of the outer m-gon between C_i and C_{i-1}. Adding the n inequalities shows that the perimeter of the inner n-gon is less than or equal to the perimeter of the outer m-gon. Thus inclusion of convex polygons necessitates growth of perimeters as well as areas.

This fact is useful in establishing inequalities among perimeters of nested convex polygons. Indeed, this was the key idea of Archimedes of Syracuse (287–212 BCE) in his approximation of π.

Archimedean inequalities

Archimedes, in *Measurement of the Circle*, found an approximation of the ratio of the circumference of a circle to its diameter by the use of inscribed and circumscribed polygons. Extending his approximation to polygons of 96 sides, he found (in modern notation)

$$3\frac{10}{71} = 3\frac{284\frac{1}{4}}{2018\frac{7}{40}} < 3\frac{284\frac{1}{4}}{2017\frac{7}{40}} < \pi < 3\frac{667\frac{1}{2}}{4673\frac{1}{2}} < 3\frac{667\frac{1}{2}}{4672\frac{1}{2}} = 3\frac{1}{7},$$

which is usually expressed by saying that π is about equal to $3\frac{1}{7}$ [Struik, 1967].

1.4   The arithmetic mean-geometric mean inequality

After the arithmetic mean the second most common mean is the *geometric mean*. For positive numbers a and b, it is given by $\sqrt{ab}$. It is often used when one wants to average ratios or factors. For example, if an investment X earns 25% in the first year (i.e., X is multiplied by a factor $a = 1.25$) and

80% in the second year (a factor $b = 1.8$), then the average annual rate of return r is $\sqrt{ab} = 1.5$ or 50% since $abX = r^2X$. If we use the arithmetic mean $(a + b)/2$ instead, we would have erroneously stated that the average rate of return is 52.5%, since the arithmetic mean of 1.25 and 1.8 is 1.525.

In this example, the arithmetic mean exceeds the geometric mean. The fact that this inequality always holds for positive numbers a and b is the celebrated *arithmetic mean-geometric mean inequality*,

$$\frac{a + b}{2} \geq \sqrt{ab},$$

with equality if and only if a and b are equal, which we henceforth abbreviate as the *AM-GM inequality*.

There are many visual proofs of this inequality. One of the simplest is the following, which uses only the Pythagorean comparison:

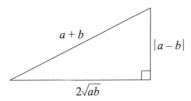

Figure 1.9.

It is easy to verify that the sum of the squares of the legs of the triangle in Figure 1.9 equals the square of the hypotenuse, and hence $a + b \geq 2\sqrt{ab}$. Division by 2 yields the AM-GM inequality.

The AM-GM inequality is a powerful problem-solving tool. We will present several examples in this and successive chapters.

What is the most important inequality, Professor Ostrowski?

The mathematician Alexander M. Ostrowski (1893–1986) made many important contributions to the theory of inequalities. Ostrowski frequently attended meetings on inequalities held at the mathematical institute in Oberwolfach, Germany. During one such meeting, a colleague of the first author relates hearing a young mathematician ask: "Professor Ostrowski, which is for you the most important inequality?" The young fellow knew of Ostrowski's many contributions to the field, and was surprised by Ostrowski's response: "The arithmetic mean-geometric mean inequality, of course!"

Application 1.1   *Given a circle of radius R, show that the inscribed rectangle with maximum area is a square.*

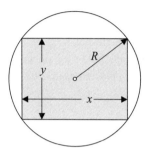

Figure 1.10.

If x and y denote the lengths of the sides of the rectangle, then $(x/2)^2 + (y/2)^2 = R^2$, and we want to maximize the area $A = xy$. Applying the AM-GM inequality, we obtain a bound on A:

$$2R^2 = \frac{x^2 + y^2}{2} \geq \sqrt{x^2 y^2} = xy = A,$$

with equality when $x = y$. Thus the inscribed rectangle of maximum area must be a square, as claimed.

Later in Challenge 4.7a we will prove a stronger statement—the inscribed *quadrilateral* with maximum area is also a square.

Application 1.2.   *Dido's problem*

The legend of Dido comes to us from the epic poem *Aeneid*, written by the Roman poet Virgil (70–19 BCE). Dido was a princess from the Phoenician city of Tyre (in present-day Lebanon). Dido fled the city after her brother murdered her husband and arrived in Africa in about 900 BCE, near the bay of Tunis. Dido decided to purchase land from the local leader, King Jarbas of Numidia, so she and her people could settle there. She paid Jarbas a sum of money for as much land as she could enclose with the hide of an ox. Virgil (John Dryden's translation) describes the scene as follows:

> At last they landed, where from far your eyes
> May view the turrets of new Carthage rise;
> There bought a space of ground, which (Byrsa call'd
> From the bull's hide) they first inclos'd, and wall'd.

To obtain as much land as possible, Dido cut the ox skin into thin strips and tied them together, as illustrated in the 17th century woodcut in Figure 1.11. This plot of land would later become the site of the city of Carthage.

Figure 1.11.

This brings us to *Dido's problem*: How should she lay the strips on the ground to enclose as much land as possible? If we assume that the ground is flat and the Mediterranean shore is straight, then the optimal solution is to lay the strips in the shape of a semicircle [Niven, 1981], which legend tells us is precisely what Dido did. We will now solve a related problem: What is the shape of the optimal rectangle in Figure 1.12?

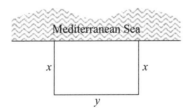

Figure 1.12.

If x and y denote the lengths of the sides of the rectangle and L the length of the strips of ox hide, then $2x + y = L$, and we want to maximize the area $A = xy$. Applying the AM-GM inequality, we have

$$A = xy = \frac{1}{2}(2xy) = \frac{1}{2}\left(\sqrt{2xy}\right)^2 \leq \frac{1}{2}\left(\frac{2x+y}{2}\right)^2 = \frac{1}{8}L^2,$$

with equality if and only if $y = 2x$. Hence the optimal rectangle is twice as wide as it is deep.

Application 1.3.   *Regiomontanus' maximum problem*

In 1471 Johannes Müller (1436–1476), called Regiomontanus after his birthplace, Königsberg, wrote a letter to Christian Roder containing the following problem:

> At what point on the earth's surface does a perpendicularly suspended rod appear longest? (that is, at what point is the visual angle a maximum?)

In his classic work *100 Great Problems of Elementary Mathematics* [Dörrie, 1965], Heinrich Dörrie writes that this problem "deserves special attention as the *first extreme problem* encountered in the history of mathematics since the days of antiquity." Our solution below is from [Maor, 1998].

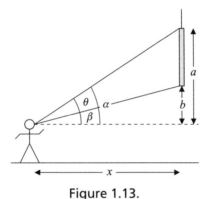

Figure 1.13.

In Figure 1.13 we see the suspended rod, whose top and bottom are a and b units, respectively, above the eye level of the observer, x units away. The task is to find x to maximize the angle θ. Let α and β denote the angles that the lines of sight to the top and bottom of the rod make, respectively, with the observer's eye level. Then

$$\cot\theta = \cot(\alpha - \beta) = \frac{\cot\alpha\cot\beta + 1}{\cot\beta - \cot\alpha}$$

$$= \frac{(x/a)(x/b) + 1}{x/b - x/a} = \frac{x}{a - b} + \frac{ab}{(a - b)x}.$$

Since the cotangent is a decreasing function for θ in the first quadrant, to maximize θ we minimize $\cot\theta$. The AM-GM inequality now yields

$$\cot\theta = \frac{x}{a - b} + \frac{ab}{(a - b)x} \geq 2\sqrt{\frac{x}{a - b} \cdot \frac{ab}{(a - b)x}} = \frac{2\sqrt{ab}}{a - b},$$

with equality if and only if $x/(a - b) = ab/(a - b)x$, or $x = \sqrt{ab}$. Thus the observer should stand at a distance equal to the geometric mean of the heights of the top and bottom of the rod above the observer's eye level.

We will encounter similar applications to other optimization problems in the next chapter when we extend the AM-GM inequality to $n \geq 3$ positive numbers.

1.5 More inequalities for means

Combining the three means we have seen so far for two positive numbers a and b, we have

$$\min(a, b) \leq \sqrt{ab} \leq \frac{a + b}{2} \leq \sqrt{\frac{a^2 + b^2}{2}} \leq \max(a, b).$$

Many more means can be inserted between $\min(a, b)$ and $\max(a, b)$. One is the *harmonic mean* of a and b, defined as the reciprocal of the arithmetic mean of the reciprocals of a and b, i.e.,

$$\frac{1}{\frac{(1/a) + (1/b)}{2}} = \frac{2ab}{a + b}.$$

The harmonic mean arises naturally in many settings. For example, if we drive 100 kilometers at 80 km/h, and another 100 kilometers at 120 km/h, then the average speed for the round trip is 96 km/h, the harmonic mean of 80 and 120.

It is easy to show that the harmonic mean of two positive numbers a and b is less than or equal to the geometric mean (simply apply the AM-GM inequality to the numbers $1/a$ and $1/b$), so the four means satisfy

$$\min(a, b) \leq \frac{2ab}{a + b} \leq \sqrt{ab} \leq \frac{a + b}{2} \leq \sqrt{\frac{a^2 + b^2}{2}} \leq \max(a, b). \quad (1.1)$$

Application 1.4. *Mengoli's inequality and the divergence of the harmonic series*

Pietro Mengoli (1625–1686) established this nice symmetric inequality [Steele, 2004]: For any $x > 1$,

$$\frac{1}{x - 1} + \frac{1}{x} + \frac{1}{x + 1} > \frac{3}{x}.$$

To verify Mengoli's inequality, we need only show that $1/(x - 1) + 1/(x + 1) > 2/x$, which is an immediate consequence of the harmonic mean-arithmetic mean inequality since the harmonic mean of $1/(x-1)$ and $1/(x+1)$ is $1/x$. Mengoli used his inequality to give the following early proof of the divergence of the harmonic series $1 + 1/2 + 1/3 + 1/4 + \cdots$. Assume the series converges to a real number H. Then write H as

$$H = 1 + \left(\frac{1}{2} + \frac{1}{3} + \frac{1}{4}\right) + \left(\frac{1}{5} + \frac{1}{6} + \frac{1}{7}\right) + \left(\frac{1}{8} + \frac{1}{9} + \frac{1}{10}\right) + \cdots .$$

Applying Mengoli's inequality yields the contradiction

$$H > 1 + \frac{3}{3} + \frac{3}{6} + \frac{3}{9} + \cdots = 1 + H.$$

The inequalities among the four means in (1.1) are illustrated using only the Pythagorean comparison in Figure 1.14 [Nelsen, 1987], where we have

$$|HM| \le |GM| \le |AM| \le |RM|.$$

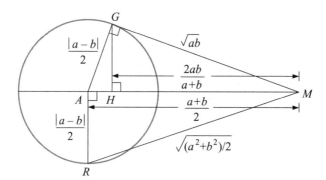

Figure 1.14.

Many other means can be inserted between $\min(a, b)$ and $\max(a, b)$. Some examples are the *contraharmonic mean* $(a^2 + b^2)/(a + b)$, the *Heronian mean* $(a + \sqrt{ab} + b)/3$, the *logarithmic mean* $(b - a)/(\ln b - \ln a)$, and the *identric mean* $(b^b/a^a)^{1/(b-a)}/e$.

Means and the apportionment of the House of Representatives

Article 1, Section 2 of the *Constitution of the United States* reads in part "Representatives and direct Taxes shall be apportioned among the several States which may be included within this Union, according to their respective Numbers,...." However, the Constitution is silent on how the apportionment is to be done, and since 1790 several different methods have been used. Normally, when one divides a state's population by a proposed House district size, the result is a non-integer that must be rounded up or down. The apportionment methods that have been used or proposed for use differ on the rounding rule. Some of the methods (and the rounding rule) are Jefferson's (always round down), Adams' (always round up), Webster's (round up or down at the arithmetic mean), Dean's (round up or down at the geometric mean), and Huntington-Hill (round up or down at the harmonic mean). The method in current use is Huntington-Hill. For further details, see [Balinski and Young, 2001].

1.6 The Ravi substitution

Consider a triangle with side lengths a, b, and c, and its inscribed circle, as shown in Figure 1.15a. Connecting the center of the circle to the vertices and the points of tangency produces three pairs of congruent right triangles, as shown in Figure 1.15b. Thus there are positive numbers x, y, and z such that $a = x + y, b = y + z, c = z + x$. This substitution is useful in establishing inequalities involving a, b, and c. This is because the only constraint on x, y, z is that they are positive, and it is often not clear how to use the triangle inequalities $a < b + c$, $b < c + a$, and $c < a + b$ in a particular problem. The substitution is known as the *Ravi substitution*.

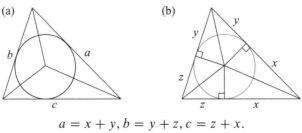

$$a = x + y, b = y + z, c = z + x.$$

Figure 1.15.

Note that $a + b - c = 2y, a - b + c = 2x$, and $-a + b + c = 2z$ and if s denotes the semiperimeter, $s = x + y + z$, so $s - a = z, s - b = x$, and $s - c = y$.

Application 1.5.   *Padoa's inequality*

This lovely inequality [Padoa, 1925], attributed to Alessandro Padoa (1868–1937) states that if a, b, c are the sides of a triangle, then

$$abc \geq (a + b - c)(b + c - a)(c + a - b). \tag{1.2}$$

Using the Ravi substitution, Padoa's inequality is equivalent to

$$(x + y)(y + z)(z + x) \geq (2x)(2y)(2z) = 8xyz.$$

This is readily established using the AM-GM inequality: $x + y \geq 2\sqrt{xy}$ and similarly for $y + z$ and $z + x$, so

$$(x + y)(y + z)(z + x) \geq 2\sqrt{xy} \cdot 2\sqrt{yz} \cdot 2\sqrt{zx} = 8xyz.$$

In Chapter 4 we will use Padoa's inequality in the form $abc \geq 8xyz$ to establish Euler's celebrated inequality relating the inradius and circumradius of a triangle.

Application 1.6.   *If a, b, c denote the sides of a triangle, show that*

$$\sqrt{a} + \sqrt{b} + \sqrt{c} \geq \sqrt{a + b - c} + \sqrt{a - b + c} + \sqrt{-a + b + c}.$$

Employing the Ravi substitution, the inequality is equivalent to

$$\sqrt{x + y} + \sqrt{y + z} + \sqrt{z + x} \geq \sqrt{2x} + \sqrt{2y} + \sqrt{2z}$$

for positive x, y, z. The inequality between the arithmetic mean and root mean square (see Section 1.1 or 1.4) applied to $\sqrt{x}$ and $\sqrt{y}$ yields $\sqrt{(x + y)/2} \geq (\sqrt{x} + \sqrt{y})/2$, or $\sqrt{x + y} \geq \sqrt{2}(\sqrt{x} + \sqrt{y})/2$ and similarly for $\sqrt{y + z}$ and $\sqrt{z + x}$. Thus

$$\sqrt{x + y} + \sqrt{y + z} + \sqrt{z + x} \geq \sqrt{2}(\sqrt{x} + \sqrt{y} + \sqrt{z})$$
$$= \sqrt{2x} + \sqrt{2y} + \sqrt{2z}.$$

1.7 Comparing graphs of functions

In Figure 1.16 [Nelsen, 1994b] we have a graph of a portion of one branch
of the rectangular hyperbola $y = 1/x$ and the line $y = 2x$ tangent to the
hyperbola at the point $(1, 1)$. Since the graph of the hyperbola is convex, it
lies on or above the tangent line for $x > 0$ (we will be precise about the
relationship between convexity and the location of tangent and secant lines
in Chapter 8). Hence for any positive x, $1/x \geq 2 - x$, or equivalently, the
sum of any positive number and its reciprocal is at least 2:

$$x > 0 \quad \Rightarrow \quad x + \frac{1}{x} \geq 2.$$

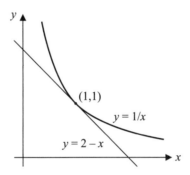

Figure 1.16.

This also follows from the AM-GM inequality applied to the numbers x
and $1/x$. We are using the *pointwise ordering* of functions, i.e., $f \leq g$ on a
set S if $f(x) \leq g(x)$ for every x in S.

The same idea can be employed to establish *Jordan's inequality* (Camille
Jordan, 1832–1922): for all x in $[0, \pi/2]$,

$$\frac{2x}{\pi} \leq \sin x \leq x.$$

In Figure 1.17 [Feng, 1996] we have a portion of the graph of the sine
function, along with graphs of the line tangent to the sine at the origin
($y = x$) and the secant line joining the origin to $(\pi/2, 1)$ ($y = 2x/\pi$).
Since the graph of the sine is concave on $[0, \pi/2]$, it must lie above the secant line and below the tangent line, yielding the desired inequality.

Care must be taken when using functional graphs to establish inequalities.
For example, in Figure 1.18 we see that $\cos x \geq 1 - x^2/2$ for $|x| \leq \pi/2$.
But the graph doesn't tell us *why* the inequality holds, unlike the situation

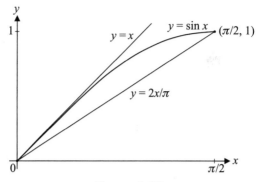

Figure 1.17.

in Figures 1.16 and 1.17, where the convexity or concavity of the functions supplies the reasons for the inequalities we see in the graphs.

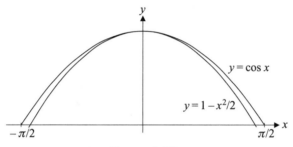

Figure 1.18.

A better way to establish $\cos x \geq 1 - x^2/2$ is to compute the length of arc and the straight line distance between the points $(\cos x, \sin x)$ and $(1, 0)$ on the unit circle as seen in Figure 1.19, and simplify.

We study the technique of this section in detail in Chapter 8.

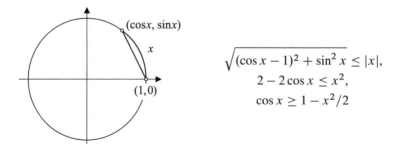

$$\sqrt{(\cos x - 1)^2 + \sin^2 x} \leq |x|,$$
$$2 - 2\cos x \leq x^2,$$
$$\cos x \geq 1 - x^2/2$$

Figure 1.19.

1.8 Challenges

1.1 In Figure 1.20, we see a non-convex quadrilateral in (a) and a skew quadrilateral in (b).

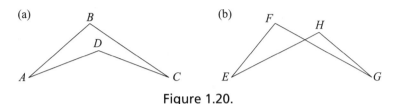

Figure 1.20.

Show that $|AD|+|DC| < |AB|+|BC|$ and $|EF|+|GH| < |EH|+|FG|$.

1.2 What basic inequalities can we derive from the following pictures?

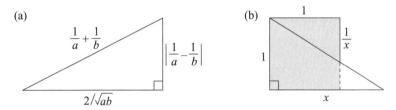

Figure 1.21.

1.3 Let a and b be the legs and c the hypotenuse of a right triangle. Create a visual proof that $a + b \leq c\sqrt{2}$.

1.4 (a) Show that the contraharmonic mean exceeds the root mean square.

   (b) Show that the difference of the arithmetic and harmonic means is the same as the difference of the contraharmonic and arithmetic means (hence the name "contraharmonic").

   (c) Show that the Heronian mean lies between the geometric and arithmetic means.

1.5 For positive numbers a, b, c, prove

$$\sqrt{a^2 - ab + b^2} + \sqrt{b^2 - bc + c^2} \geq \sqrt{a^2 + ac + c^2}.$$

1.6 Use graphs of functions to establish *Bernoulli's inequality* (Johann Bernoulli, 1667–1748): For $x > 0$ and $r > 1$, $x^r - 1 \geq r(x - 1)$.

1.7 Prove: If $x > 0$, then $\sqrt[x]{x} \leq \sqrt[e]{e}$. (Hint: consider the graphs of $y = e^{x/e}$ and its tangent line at (e, e).)

1.8 Use the AM-GM inequality (rather than calculus) to find the best design for an open box with one partition, as shown in Figure 1.22. "Best" can mean either maximum volume for a fixed amount of material, or minimum amount of material for a fixed volume. (Ignore the thickness of the material used to construct the box.)

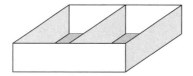

Figure 1.22.

1.9 Let a, b, c denote the lengths of the sides of a triangle, and s its semiperimeter. Prove the following inequalities. (Hint: the Ravi substitution may help.)

(a) $\dfrac{1}{s-a} + \dfrac{1}{s-b} + \dfrac{1}{s-c} \geq \dfrac{9}{s}$.

(b) $\sqrt{s} \leq \sqrt{s-a} + \sqrt{s-b} + \sqrt{s-c} \leq \sqrt{3s}$.

1.10 Let a, b, c denote the sides of a triangle. Show that

$$a + b - (2 - \sqrt{2 - 2\cos C})\max(a, b) \leq c$$
$$\leq a + b - (2 - \sqrt{2 - 2\cos C})\min(a, b).$$

1.11 Show that the sequence $\{a_n\}$ with $a_n = 2^n \sin(\pi/2^n)$, $n = 1, 2, \ldots$ is increasing and bounded.

CHAPTER 2

Representing positive numbers as areas or volumes

We now extend the inclusion principle from the first chapter by representing a positive number as the area or the volume of an object, and showing that one object is included in the other. We begin with area representations of numbers, primarily by rectangles and triangles.

The inclusion principle will be applied in two ways. Suppose we wish to establish $x \leq y$ and we have found a region A with area x and a region B with area y. Then $x \leq y$ if either (a) A fits inside B, or (b) pieces of B cover A with possible overlap of some pieces. In Figure 2.1 we illustrate both methods by again establishing the AM-GM inequality for positive numbers a and b:

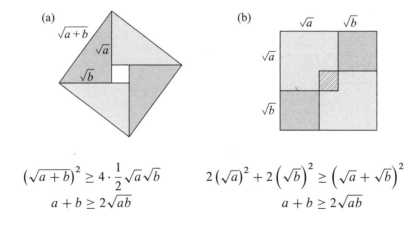

(a)
$$\left(\sqrt{a+b}\right)^2 \geq 4 \cdot \frac{1}{2}\sqrt{a}\sqrt{b}$$
$$a + b \geq 2\sqrt{ab}$$

(b)
$$2\left(\sqrt{a}\right)^2 + 2\left(\sqrt{b}\right)^2 \geq \left(\sqrt{a} + \sqrt{b}\right)^2$$
$$a + b \geq 2\sqrt{ab}$$

Figure 2.1.

In Figure 2.1a, each triangle, similar to the one in Figure 1.3, has area $\sqrt{a}\sqrt{b}/2$, and four of them fit inside a square with side length $\sqrt{a+b}$. In

Figure 2.1b, two squares with side length $\sqrt{a}$ and two with side length $\sqrt{b}$ cover a square with sides $\sqrt{a} + \sqrt{b}$.

2.1   Three examples

Example 2.1. The first example illustrates the cross-multiplication criterion for the inequality between positive fractions:

For a, b, c, d positive, $\dfrac{a}{b} \leq \dfrac{c}{d}$ if and only if $ad \leq bc$.

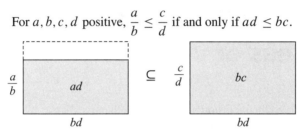

Figure 2.2.

Example 2.2. The AM-GM inequality can be illustrated using right triangles whose areas are the positive numbers a and b [Beckenbach and Bellman, 1961]:

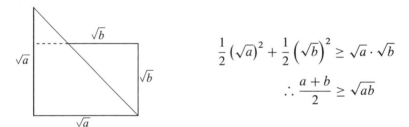

$$\frac{1}{2}\left(\sqrt{a}\right)^2 + \frac{1}{2}\left(\sqrt{b}\right)^2 \geq \sqrt{a} \cdot \sqrt{b}$$

$$\therefore \frac{a+b}{2} \geq \sqrt{ab}$$

Figure 2.3.

Heron's approximation of square roots

In his book *Metrika*, Heron of Alexandria (second half of the first century) presented a method for approximating the square root $\sqrt{n}$ of an integer n: if $n = ab$, then $\sqrt{n}$ is approximately $(a + b)/2$, and the approximation is better when a is close to b. We can see this as a consequence of the AM-GM inequality: $\sqrt{n} = \sqrt{ab} \leq (a + b)/2$. In fact, if x_1 is an approximation to $\sqrt{n}$, then $x_2 = [x_1 + (n/x_1)]/2$ is a better one, and so on. Newton's method is based on this idea.

Example 2.3. For any four positive numbers a, b, c, d with $a \leq b, c \leq d$, we have

$$ad + bc \leq ac + bd.$$

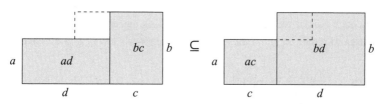

Figure 2.4.

The inequality also holds if $a \geq b$ and $c \geq d$ with equality if and only if $a = b$ or $c = d$. If $a = c = \sqrt{x}$ and $b = d = \sqrt{y}$, we have yet another proof of the AM-GM inequality. If we set $c = a^2$ and $d = b^2$, then we have $a^2 b + ab^2 \leq a^3 + b^3$. An inequality involving cubes suggests a three-dimensional picture. We can extend the idea behind Figure 2.4 by using volumes of boxes. In Figure 2.5 we see that $a \leq b$ implies $a^2 b + ab^2 \leq a^3 + b^3$:

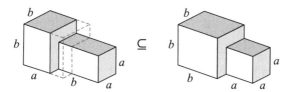

Figure 2.5.

2.2 Chebyshev's inequality

The result in Example 2.3 can now be used to establish *Chebyshev's inequality* (Pafnuty Lvovich Chebyshev, 1821–1894):

Theorem 2.1. *For any $n \geq 2$, let $0 < x_1 \leq x_2 \leq \cdots \leq x_n$. Then*

(i) if $0 < y_1 \leq y_2 \leq \cdots \leq y_n$, then $\sum_{i=1}^{n} x_i \sum_{i=1}^{n} y_i \leq n \sum_{i=1}^{n} x_i y_i$,

$$(2.1a)$$

(ii) if $y_1 \geq y_2 \geq \cdots \geq y_n > 0$, then $\sum_{i=1}^{n} x_i \sum_{i=1}^{n} y_i \geq n \sum_{i=1}^{n} x_i y_i$,

$$(2.1b)$$

with equality in each if and only if all the x_is are equal, or all the y_is are equal.

Proof. For (i), we use the result in Example 2.3 with $a = x_i, b = x_j$, $c = y_i, d = y_j$, so $x_i y_j + x_j y_i \leq x_i y_i + x_j y_j$. Then, as illustrated in Figure 2.6, each pair of rectangles with solid shading has area less than or equal to the area of a pair of rectangles with total area $x_i y_i + x_j y_j$, and hence

$$(x_1 + x_2 + \cdots + x_n)(y_1 + y_2 + \cdots + y_n) \leq n(x_1 y_1 + x_2 y_2 + \cdots + x_n y_n).$$

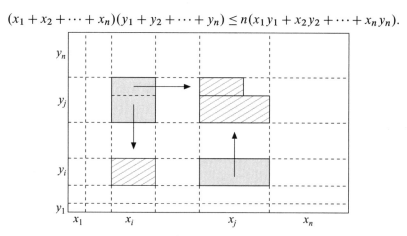

Figure 2.6.

For (ii), we use the result in Example 2.3 with $a = x_i, b = x_j, c = y_j, d = y_i$, so $x_i y_j + x_j y_i \geq x_i y_i + x_j y_j$. Hence the inequality is reversed, and

$$(x_1 + x_2 + \cdots + x_n)(y_1 + y_2 + \cdots + y_n) \geq n(x_1 y_1 + x_2 y_2 + \cdots + x_n y_n),$$

which concludes the proof. When $y_i = 1/x_i$, we have

$$(x_1 + x_2 + \cdots + x_n)\left(\frac{1}{x_1} + \frac{1}{x_2} + \cdots + \frac{1}{x_n}\right) \geq n^2, \qquad (2.2)$$

with equality if and only if all the x_is are equal, which we now use in two applications.

Application 2.1. *Nesbitt's inequality* [Nesbitt, 1903]. This inequality, a staple of mathematical competitions, states that if a, b, c are positive, then

$$\frac{a}{b+c} + \frac{b}{c+a} + \frac{c}{a+b} \geq \frac{3}{2}.$$

This inequality follows directly from inequality (2.2) applied to $a+b, b+c$, and $c+a$:

$$\frac{a}{b+c} + \frac{b}{c+a} + \frac{c}{a+b}$$

$$= \left(\frac{a}{b+c}+1\right) + \left(\frac{b}{c+a}+1\right) + \left(\frac{c}{a+b}+1\right) - 3$$

$$= (a+b+c)\left(\frac{1}{a+b} + \frac{1}{b+c} + \frac{1}{c+a}\right) - 3$$

$$= \frac{1}{2}\left[(a+b) + (b+c) + (c+a)\right]\left(\frac{1}{a+b} + \frac{1}{b+c} + \frac{1}{c+a}\right) - 3$$

$$\geq \left(\frac{1}{2}\cdot 9\right) - 3 = \frac{3}{2}.$$

Application 2.2.   *Voicu's inequality* [Voicu, 1981]. Let α, β, γ denote the angles that an interior diagonal of a rectangular box makes with the edges, as shown in Figure 2.7. Show that

$$\tan\alpha \tan\beta \tan\gamma \geq 2\sqrt{2},$$

with equality if and only if the box is a cube.

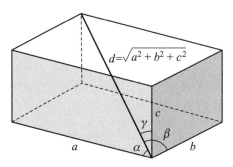

Figure 2.7.

From the figure, we have

$$\cos^2\alpha + \cos^2\beta + \cos^2\gamma$$

$$= \frac{a^2}{a^2+b^2+c^2} + \frac{b^2}{a^2+b^2+c^2} + \frac{c^2}{a^2+b^2+c^2} = 1.$$

Let $x = \cos\alpha$, $y = \cos\beta$, $z = \cos\gamma$, and apply inequality (2.2) for three numbers to x^2, y^2, and z^2:

$$1 + \tan^2\alpha \tan^2\beta \tan^2\gamma$$

$$= 1 + \frac{1-x^2}{x^2} \cdot \frac{1-y^2}{y^2} \cdot \frac{1-z^2}{z^2} = \frac{x^2y^2 + y^2z^2 + z^2x^2}{x^2y^2z^2}$$

$$= \frac{1}{x^2} + \frac{1}{y^2} + \frac{1}{z^2} = (x^2 + y^2 + z^2)\left(\frac{1}{x^2} + \frac{1}{y^2} + \frac{1}{z^2}\right) \geq 9,$$

and consequently $\tan\alpha \tan\beta \tan\gamma \geq 2\sqrt{2}$, with equality if and only if each tangent is $\sqrt{2}$, i.e., when the box is a cube.

2.3   The AM-GM inequality for three numbers

To establish the AM-GM inequality $\sqrt[3]{xyz} \leq (x + y + z)/3$ for positive numbers x, y, z we first make a change of variables $x = a^3$, $y = b^3$, $z = c^3$, so we can work with the inequality

$$3abc \leq a^3 + b^3 + c^3.$$

We first prove a preliminary lemma [Alsina, 2000], which is of interest in its own right:

Lemma 2.1. *For all a, b, c $\geq$ 0, $ab + bc + ac \leq a^2 + b^2 + c^2$.*

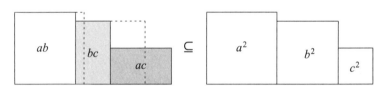

Figure 2.8.

Without loss of generality, we can assume $a \geq b \geq c$, and Figure 2.8 illustrates the inequality in this case. This can also be established by applying the AM-GM inequality to the pairs a^2 and b^2, b^2 and c^2, c^2 and a^2.

Application 2.3.   *An inequality of Guba* [Guba, 1977]

Consider a rectangular box with edge lengths a, b, c so the areas of the faces are $K_1 = ab$, $K_2 = bc$, and $K_3 = ac$. The volume is $V = abc$ and the interior diagonal $d = \sqrt{a^2 + b^2 + c^2}$. Then Guba's inequality states that

$$K_1^2 + K_2^2 + K_3^2 \geq \sqrt{3}Vd.$$

Using Lemma 2.1 we see that

$$(a + b + c)^2 = a^2 + b^2 + c^2 + 2(ab + bc + ac)$$
$$\geq 3(ab + bc + ac).$$

Thus

$$\left(K_1^2 + K_2^2 + K_3^2\right)^2 \geq 3\left(K_1^2 K_2^2 + K_2^2 K_3^2 + K_1^2 K_3^2\right)$$
$$= 3a^2 b^2 c^2 \left(a^2 + b^2 + c^2\right) = 3V^2 d^2,$$

which proves the inequality.

Theorem 2.2. *For all a, b, c $\geq$ 0, $3abc \leq a^3 + b^3 + c^3$.*

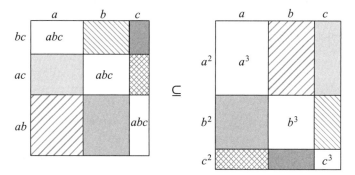

Figure 2.9.

Proof. [Alsina, 2000]. In Figure 2.9, rectangles with the same shading have the same area, and since from Lemma 2.1 the height of the left-side rectangle is less than or equal to the height of the right-side rectangle, the three white rectangles on the left have area $3abc$ less than or equal to the area of the three white rectangles on the right, $a^3 + b^3 + c^3$.

Since the above version of the AM-GM inequality for three numbers involves cubes, we can illustrate it with a three-dimensional picture. In Figure 2.10, we see that a box with side lengths $a \geq b \geq c$ fits inside the union of three right pyramids whose bases are squares with side lengths a, b, c, and whose altitudes are also a, b, c, respectively, so

$$abc \leq \frac{1}{3}a^2 \cdot a + \frac{1}{3}b^2 \cdot b + \frac{1}{3}c^2 \cdot c.$$

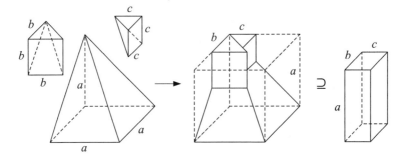

Figure 2.10.

Application 2.4 *Among all triangles with a fixed perimeter, the equilateral triangle has the greatest area*

Consider a triangle with side lengths a, b, c, area K, and *semi-perimeter* $s = (a + b + c)/2$. By the AM-GM inequality, we have

$$\sqrt[3]{(s - a)(s - b)(s - c)} \le \frac{(s - a) + (s - b) + (s - c)}{3} = \frac{s}{3},$$

or $(s - a)(s - b)(s - c) \le s^3/27$, with equality if and only if $a = b = c$. The classic formula of Heron of Alexandria (c. 10–75), which we will prove in Application 4.1, states that $K = \sqrt{s(s - a)(s - b)(s - c)}$ and hence

$$K^2 = s(s - a)(s - b)(s - c) = s\left[\sqrt[3]{(s - a)(s - b)(s - c)}\right]^3 \le \frac{s^4}{27},$$

or $K \le \sqrt{3}s^2/9$. Since the perimeter is fixed, so is s, and hence the area is largest when we have equality, or when the triangle is equilateral.

Many optimizations problems typically encountered in single variable calculus courses can be solved using the AM-GM inequality for $n \ge 3$ numbers. The following Application illustrates the technique; and similar problems can be found in the Challenges.

Application 2.5 *Find the dimensions and volume of the right circular cylinder with maximum volume that can be inscribed in a right circular cone with base radius R and altitude H.*

Let r and h denote, respectively, the base radius and altitude of the inscribed cylinder. If we introduce an axis system in the plane containing the axis of the cone, then the point (r, h) lies on the line $(x/R) + (y/H) = 1$, so $(r/R) + (h/H) = 1$, as seen in Figure 2.11.

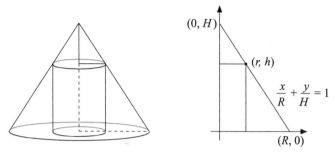

Figure 2.11.

Since the volume of the cylinder is $V = \pi r^2 h$, we have

$$V = \pi r^2 h = 4\pi R^2 H \cdot \frac{r}{2R} \cdot \frac{r}{2R} \cdot \frac{h}{H}$$

$$\leq 4\pi R^2 H \cdot \left(\frac{(r/2R) + (r/2R) + (h/H)}{3} \right)^3$$

$$= 4\pi R^2 H \cdot \frac{1}{27} = \frac{4}{9} \cdot \frac{1}{3} \pi R^2 H,$$

with equality if and only if $r/(2R) = h/H$. Since $(r/R) + (h/H) = 1$, it follows that the cylinder has maximum volume when $r = 2R/3$ and $h = H/3$, and that its volume is $4/9$ the volume of the given cone.

Application 2.6 *Extremes of sums and products of tangents*

Consider a triangle with angles measuring α, β, and γ. We will show that

$$\tan \frac{\alpha}{2} \tan \frac{\beta}{2} \tan \frac{\gamma}{2} \leq \frac{\sqrt{3}}{9} \quad \text{and} \quad \tan \frac{\alpha}{2} + \tan \frac{\beta}{2} + \tan \frac{\gamma}{2} \geq \sqrt{3}.$$

Since $\alpha + \beta + \gamma = \pi$, $\tan((\alpha + \beta)/2) = \cot(\gamma/2)$, so

$$\frac{\tan(\alpha/2) + \tan(\beta/2)}{1 - \tan(\alpha/2)\tan(\beta/2)} = \frac{1}{\tan(\gamma/2)},$$

and hence

$$\tan(\alpha/2)\tan(\beta/2) + \tan(\beta/2)\tan(\gamma/2) + \tan(\gamma/2)\tan(\alpha/2) = 1.$$

Now let $x = \tan(\alpha/2)$, $y = \tan(\beta/2)$, and $z = \tan(\gamma/2)$. Our task is to show that $xyz \leq \sqrt{3}/9$ and $x + y + z \geq \sqrt{3}$ when $xy + yz + zx = 1$.

Applying the AM-GM inequality to the three numbers xy, yz, and zx yields

$$\frac{1}{3} = \frac{xy + yz + zx}{3} \geq \sqrt[3]{xy \cdot yz \cdot zx} = (xyz)^{2/3},$$

from which $xyz \leq \sqrt{3}/9$ follows. Lemma 2.1 yields

$$(x+y+z)^2 = x^2+y^2+z^2+2(xy+yz+zx) \geq 3(xy+yz+zx), \quad (2.3)$$

and since $xy + yz + zx = 1$, we have $x + y + z \geq \sqrt{3}$, as required. We have equality in both inequalities if and only if $x = y = z = \sqrt{3}/3$, i.e., when the triangle is equilateral. We will encounter similar inequalities for sines and cosines in Application 7.1.

Application 2.7 *Newton's inequality*

Given n real numbers $a_1, a_2, \ldots, a_n$ and a fixed $i, 0 \leq i \leq n$, the ith *elementary symmetric function* σ_i is defined to be the coefficient of x^{n-i} in the expansion of $(x + a_1)(x + a_2) \cdots (x + a_n)$. Associated with each σ_i is the ith *elementary symmetric mean* S_i, defined as $S_i = \sigma_i / \binom{n}{i}$. For example, when $n = 3$, we have $S_0 = 1, S_1 = (a_1 + a_2 + a_3)/3, S_2 = (a_1a_2 + a_2a_3 + a_3a_1)/3$, and $S_3 = a_1a_2a_3$. *Newton's inequality* (Isaac Newton, 1642–1727) establishes that $S_{i-1}S_{i+1} \leq S_i^2$. We now give a short proof of the special case $n = 3, i = 2$, and $a_1, a_2, a_3 \geq 0$:

$$\frac{a_1 + a_2 + a_3}{3} \cdot a_1a_2a_3 \leq \left(\frac{a_1a_2 + a_2a_3 + a_3a_1}{3} \right)^2 .$$

If we set $x = a_1a_2, y = a_2a_3$, and $z = a_3a_1$, the inequality is equivalent to

$$3(xy + yz + zx) \leq (x + y + z)^2,$$

which is (2.3).

The Airline Inequality.

To avoid long lines, extra fees, and wasted time checking and retrieving luggage at airports, many travelers chose to have only carry-on luggage. However, most airlines have size and weight limits for carry-on luggage, typically restricting the dimensions x, y, z of a carry-on bag to $x + y + z \leq 45$ inches. The AM-GM inequality shows that the shape of such a suitcase with maximum volume xyz is a cube 15 inches on a side. However, such a bag rarely fits into the overhead bin.

We can extend the AM-GM inequality to four positive numbers a, b, c, d, using the two-number version twice:

$$\frac{a+b+c+d}{4} = \frac{1}{2}\left(\frac{a+b}{2} + \frac{c+d}{2}\right) \geq \frac{1}{2}\left(\sqrt{ab} + \sqrt{cd}\right)$$

$$\geq \sqrt{\sqrt{ab}\sqrt{cd}} = \sqrt[4]{abcd}.$$

We can similarly extend the inequality to n numbers when n is a power of 2 (see Challenge 2.2). To extend the inequality to all positive integers n, we will first prove the following inequality.

2.4 Guha's inequality

This inequality states that if $a \geq 0$, $p \geq q > 0$ and $x \geq y > 0$, then

$$(px + y + a)(x + qy + a) \geq [(p+1)x + a]\,[(q+1)y + a].$$

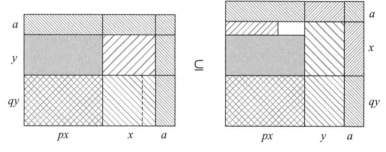

Figure 2.12.

Algebraically, the inequality is equivalent to $(px - qy)(x - y) \geq 0$, which says that the area of the small white rectangle on the right is nonnegative. Consequently, we have equality if and only if $x = y$. Guha used this inequality in an elementary but clever proof of the AM-GM inequality for n numbers [Guha, 1967] that we present in the next section.

2.5 The AM-GM inequality for n numbers

The AM-GM inequality for n positive numbers $x_1, x_2, \ldots, x_n$ states that $G_n \leq A_n$, where $A_n = (x_1 + x_2 + \cdots + x_n)/n$ and $G_n = (x_1 x_2 \cdots x_n)^{1/n}$. Without loss of generality, assume $x_1 \leq x_2 \leq \cdots \leq x_n$. Repeated application of Guha's inequality is all that is required to establish the AM-GM inequality for n numbers. We illustrate for $n = 4$. The procedure is identical for any $n \geq 2$.

$$
\begin{aligned}
(4A_4)^4 &= (a+b+c+d)(a+b+c+d)(a+b+c+d)(a+b+c+d) \\
&\geq (2a+c+d)(2b+c+d)(a+b+c+d)(a+b+c+d) \\
&\geq (3a+d)(2b+c+d)(b+2c+d)(a+b+c+d) \\
&\geq (4a)(2b+c+d)(b+2c+d)(b+c+2d) \\
&\geq (4a)(3b+d)(3c+d)(b+c+2d) \\
&\geq (4a)(4b)(3c+d)(c+3d) \\
&\geq (4a)(4b)(4c)(4d) = (4G_4)^4,
\end{aligned}
$$

with equality if and only if $a = b = c = d$.

We now use the AM-GM inequality for four numbers to prove

Theorem 2.3. *Among all triangles that can be inscribed in a given circle, the equilateral triangle has the largest area.*

Proof. We first note that given any non-isosceles triangle inscribed in the circle, there is an isosceles triangle that has a greater height and thus larger area, as illustrated in Figure 2.13a. Hence we need only consider isosceles triangles.

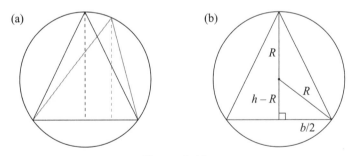

Figure 2.13.

Let R denote the radius of the circle, and consider an inscribed isosceles triangle with base length b and height h, as illustrated in Figure 2.13b. We can assume $h \geq R$. Then $(b/2)^2 + (h-R)^2 = R^2$, or $(b/2)^2 = h(2R-h)$.

If K denotes the area of the triangle, then

$$K^2 = (bh/2)^2 = h^3(2R - h) = 27 \left(\frac{h}{3}\right)^3 (2R - h)$$

$$\leq 27 \left[\frac{1}{4}\left(\frac{h}{3} + \frac{h}{3} + \frac{h}{3} + (2R - h)\right)\right]^4 = 27 \left(\frac{R}{2}\right)^4 = \frac{27}{16}R^4,$$

or $K \leq 3\sqrt{3}R^2/4$, with equality if and only if $h = 3R/2$, i.e., when the triangle is equilateral.

Corollary 2.1. *If a triangle with area K is inscribed in a circle of radius R, then $K \leq 3\sqrt{3}R^2/4$.*

2.6 The HM-AM-GM-RMS inequality for n numbers

We can easily extend the AM-GM inequality to $H_n \leq G_n \leq A_n \leq R_n$, where H_n denotes the *harmonic mean* (HM) of $x_1, x_2, \ldots, x_n$,

$$H_n = \frac{n}{\dfrac{1}{x_1} + \dfrac{1}{x_2} + \cdots + \dfrac{1}{x_n}};$$

and R_n denotes the *root-mean-square* (RMS) of $x_1, x_2, \ldots, x_n$,

$$R_n = \sqrt{\frac{x_1^2 + x_2^2 + \cdots + x_n^2}{n}}.$$

Applying the AM-GM inequality to $1/x_1, 1/x_2, \ldots, 1/x_n$ yields

$$\frac{1}{n}\left(\frac{1}{x_1} + \frac{1}{x_2} + \cdots + \frac{1}{x_n}\right) \geq \sqrt[n]{\frac{1}{x_1} \cdot \frac{1}{x_2} \cdots \frac{1}{x_n}},$$

or $H_n \leq G_n$. Without loss of generality we may now assume $x_1 \leq x_2 \leq \cdots \leq x_n$, let $y_i = x_i$, and apply (2.1a) to get

$$(x_1 + x_2 + \cdots + x_n)^2 \leq n\left(x_1^2 + x_2^2 + \cdots + x_n^2\right),$$

or $A_n \leq R_n$. Thus we have proven

Theorem 2.4. *For any positive numbers $x_1, x_2, \ldots, x_n$, we have*

$$\frac{n}{\dfrac{1}{x_1} + \dfrac{1}{x_2} + \cdots + \dfrac{1}{x_n}} \leq \sqrt[n]{x_1 x_2 \cdots x_n} \leq \frac{x_1 + x_2 + \cdots + x_n}{n}$$

$$\leq \sqrt{\frac{x_1^2 + x_2^2 + \cdots + x_n^2}{n}}.$$

2.7 The mediant property and Simpson's paradox

The *mediant* of two fractions a/b and c/d, where $b > 0$ and $d > 0$, is the fraction $(a + c)/(b + d)$. If $a/b < c/d$, then

$$\frac{a}{b} < \frac{a + c}{b + d} < \frac{c}{d},$$

i.e., the fraction that results from adding numerators and adding denominators always lies between the given fractions. This inequality is known as the *mediant property*, which we illustrate (for $a > 0$ and $c > 0$) the equivalent inequality

$$\frac{a}{b} < \frac{a}{b + d} + \frac{c}{b + d} < \frac{c}{d}$$

in Figure 2.14:

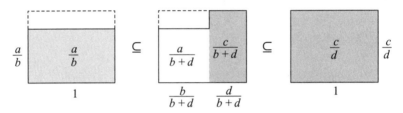

Figure 2.14.

A fraction can also represent the slope of a line segment, which leads to a second illustration via the parallelogram law for vector addition. Here we see that the slope $(a + c)/(b + d)$ of the diagonal of the parallelogram lies between the slopes a/b and c/d of the two sides of the parallelogram emanating from the origin:

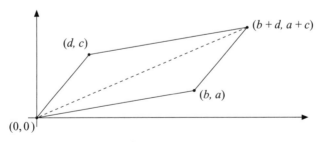

Figure 2.15.

The arithmetic mean is *monotonic*, in the sense that if $x < X$ and $y < Y$, then $(x + y)/2 < (X + Y)/2$. The mediant is *not* monotonic, which leads to some surprising results, as we illustrate in Application 2.8 below.

Chuquet, Poincaré, and the mediant property

Nicolas Chuquet (circa 1450–1500) discussed the mediant in his book *Le Triparty en la science des Nombres*, published in 1484. Chuquet referred to this calculation as *le règle des nombres moyens*, the rule of mean numbers. Several centuries later the well-known mathematician Henri Poincaré (1854–1912) said [Arnol'd, 2006] "If you want to teach fractions either you divide cakes, even in a virtual way, or you bring an apple to the classroom. In all other cases students will continue adding numerators and adding denominators."

Application 2.8 *Simpson's paradox*

Simpson's Paradox refers to [Moore, 2004] "an association or comparison that holds for all of several groups can reverse direction when the data are combined to form a single group." The comparisons are often expressed as fractions, i.e., a/b represents a successes in a group of size b. We may have

$$\frac{a}{b} < \frac{A}{B}, \quad \frac{c}{d} < \frac{C}{D}, \quad \text{but yet} \quad \frac{a+c}{b+d} > \frac{A+C}{B+D}.$$

This is sometimes called a *Simpson's reversal of inequalities*. For example [Stanford]:

> Suppose that a University is trying to discriminate in favor of women when hiring staff. It advertises positions in the Department of History and in the Department of Geography, and only those departments. Five men apply for the positions in History and one is hired, and eight women apply and two are hired. The success rate for men is twenty percent, and the success rate for women is twenty-five percent. The History Department has favored women over men. In the Geography Department eight men apply and six are hired, and five women apply and four are hired. The success rate for men is seventy-five percent and for women it is eighty percent. The Geography Department has favored women over men. Yet across the University as a whole 13 men and 13 women applied for jobs, and 7 men and 6 women were hired. The success rate for male applicants is greater than the success rate for female applicants.

Since the overall success rates $(a + c)/(b + d)$ and $(A + C)/(B + D)$ are the mediants of the rates a/b, c/d and A/B, C/D, respectively, vector-addition parallelograms help explain the reversal, as illustrated in Figure 2.16 [Kocik, 2001]:

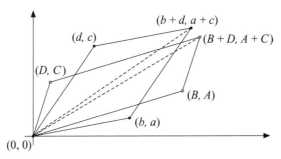

Figure 2.16.

Here we see that it is possible to have $a/b < A/B$ and $c/d < C/D$, yet $(a + c)/(b + d) > (A + C)/(B + D)$.

For another example of representing fractions as slopes of line segments, see Challenge 2.21.

2.8   Chebyshev's inequality revisited

Chebyshev's inequality (2.1a) from Section 2.2 has several nice applications, which we now examine. First we present another proof of (2.1a) by area inclusion in Figure 2.17.

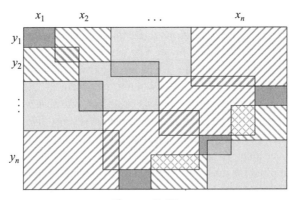

Figure 2.17.

In the figure, n copies of rectangles with areas $x_1 y_1, x_2 y_2, \ldots, x_n y_n$ are arranged to cover with some overlap a rectangle with dimensions $x_1 + x_2 + \cdots + x_n$ by $y_1 + y_2 + \cdots + y_n$, so

$$n(x_1 y_1 + x_2 y_2 + \cdots + x_n y_n) \geq (x_1 + x_2 + \cdots + x_n)(y_1 + y_2 + \cdots + y_n).$$

$$(2.4)$$

Application 2.9 *The mean of the squares exceeds the square of the mean*

Consider a set $\{a_1, a_2, \ldots, a_n\}$ of real numbers, and let $x_i = y_i = |a_i|$ in Chebyshev's inequality (2.4). Then

$$n \left(a_1^2 + a_2^2 + \cdots + a_n^2 \right) \geq (|a_1| + |a_2| + \cdots + |a_n|)^2$$
$$\geq (a_1 + a_2 + \cdots + a_n)^2,$$

or

$$\frac{1}{n} \left(a_1^2 + a_2^2 + \cdots + a_n^2 \right) \geq \left(\frac{a_1 + a_2 + \cdots + a_n}{n} \right)^2.$$

This inequality, although identical to the arithmetic mean-root mean square inequality for n positive numbers, is stronger, since it applies to all real $a_1, a_2, \ldots, a_n$. In statistics, it is equivalent to the fact that the sample variance of a data set is nonnegative.

Application 2.10 *Power means*

If we employ (2.3) twice, first with $y_i = x_i^2$ and then with $y_i = x_i$, we have

$$n^2 \left(x_1^3 + x_2^3 + \cdots + x_n^3 \right) \geq (x_1 + x_2 + \cdots + x_n) \cdot n \left(x_1^2 + x_2^2 + \cdots + x_n^2 \right)$$
$$\geq (x_1 + x_2 + \cdots + x_n)^3,$$

or

$$\frac{x_1 + x_2 + \cdots + x_n}{n} \leq \sqrt[3]{\frac{x_1^3 + x_2^3 + \cdots + x_n^3}{n}}.$$

The expression on the right is known as a *power mean*, and the process can be repeated to establish the inequality between the arithmetic mean and the

pth power mean of positive numbers x_i, for any integer $p \geq 2$:

$$\frac{x_1 + x_2 + \cdots + x_n}{n} \leq \sqrt[p]{\frac{x_1^p + x_2^p + \cdots + x_n^p}{n}}.$$

Using Bernoulli's inequality, we can establish the *power mean inequality*:

Theorem 2.5. *Let* $x_1, x_2, \ldots, x_n, p, q$ *be positive real numbers, with* $p \leq q$.
Then

$$\sqrt[p]{\frac{x_1^p + x_2^p + \cdots + x_n^p}{n}} \leq \sqrt[q]{\frac{x_1^q + x_2^q + \cdots + x_n^q}{n}}.$$

Proof. Let $A = \left((\sum x_i^p)/n \right)^{1/p}$ and $B = \left((\sum x_i^q)/n \right)^{1/q}$ (all sums are for i from 1 to n). Then

$$\frac{B}{A} = \left(\frac{1}{n} \left[\sum \left(\frac{x_i}{A} \right)^q \right] \right)^{1/q} = \left[\frac{1}{n} \left(\sum \left[\left(\frac{x_i}{A} \right)^p \right]^{q/p} \right) \right]^{1/q}.$$

We now apply Bernouilli's inequality (see Challenge 1.6) $y^r \geq 1 + r(y - 1)$ with $y = x_i/A > 0$ and $r = q/p$ to obtain

$$\frac{B}{A} \geq \left(\frac{1}{n} \left[\sum \left(1 + \frac{q}{p} \left[\left(\frac{x_i}{A} \right) - 1 \right] \right) \right] \right) = 1.$$

Power means for three numbers and human poverty

The *Human Poverty Index* HPI-1 is a composite index introduced by the United Nations to measure deprivation in three basic dimensions of human development: a long and healthy life, knowledge, and a decent standard of living. It is calculated as a power mean:

$$\text{HPI} - 1 = \left[(P_1^3 + P_2^3 + P_3^3)/3 \right]^{1/3},$$

where P_1 is the probability at birth of not surviving to age 40 (times 100), P_2 is the adult illiteracy rate, and P_3 is the unweighted average of population without sustainable access to an improved water source and children under weight for age. According to [Human Development Report, 2006], the power 3 is used "to give additional but not overwhelming weight to areas of more acute deprivation."

Power means, Piet Hein, and the superellipse

For positive a, b, and p, consider points (x, y) in the plane such that the pth power mean of $|x/a|$ and $|y/b|$ is constant, i.e., $\sqrt[p]{(|x/a|^p + |y/b|^p)/2} = k$. For $k = 2^{-1/p}$, this simplifies to $|x/a|^p + |y/b|^p = 1$. When $p = 2$, the graph of this equation is an ellipse, but for $p > 2$, it is called a *superellipse* (or *Lamé curve*). Superellipses were popularized by the Danish mathematician, inventor, and poet Piet Hein (1905–1996), although they were first studied in 1818 by the French mathematician Gabriel Lamé (1795–1870). Hein used a superellipse with $p = 2.5$ and $a/b = 6/5$ in his 1959 design for a traffic roundabout for the Sergels Torg, a city square in Stockholm, Sweden. The Azteca Stadium, built for the 1968 Olympic Games in Mexico City, also has the shape of a superellipse.

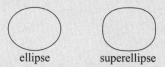

ellipse   superellipse

Hein created many mathematical recreations such as the Soma cube and the games of Hex and TacTix. He also wrote thousands of short aphoristic poems called *grooks* (*gruk* in Danish); here is an example (appropriate for inclusion in a book about mathematics!) entitled "Problems":

> Problems worthy
> of attack,
> prove their worth
> by hitting back.

2.9 Schur's inequality

The classical inequality of Issai Schur (1875–1941) states that for nonnegative x, y, z and positive r, we have

$$x^r(x - y)(x - z) + y^r(y - z)(y - x) + z^r(z - x)(z - y) \geq 0,$$

with equality if $x = y = z$ or if two of the variables are equal and the third zero. Given the symmetry in the inequality, we may assume that $x \geq y \geq z$, and rewrite it as

$$x^r(x - y)(x - z) + z^r(x - z)(y - z) \geq y^r(x - y)(y - z),$$

and interpret each term as the volume of a box. Since $x^r \geq y^r$ and $x - z \geq y - z$, we have $x^r(x-y)(x-z) \geq y^r(x-y)(y-z)$ and since $z^r(x-z)(y-z) \geq 0$, the inequality holds. In Figure 2.18, we see boxes whose volumes are the terms in the inequality. The box whose volume is the term on the right is shaded gray, and is contained in the union of the other two boxes. The rth powers of x, y, z can be replaced by any nondecreasing positive function of the variables.

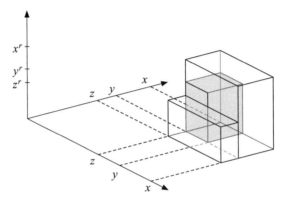

Figure 2.18.

Application 2.11   *Let a, b, c denote the lengths of the sides of a triangle, and s its semiperimeter. Prove* $abcs \geq a^3(s-a) + b^3(s-b) + c^3(s-c)$.

Bring all the terms to the left side and multiply by two. Then write $2abcs = a^2bc + ab^2c + abc^2$, and observe that

$$a^2bc - 2a^3(s-a) = a^2bc - a^3(b+c-a) = a^2(a-b)(a-c),$$

and similarly

$$ab^2c - 2b^3(s-b) = b^2(b-c)(b-a),$$
$$abc^2 - 2c^3(s-c) = c^2(c-a)(c-b).$$

Hence the given inequality is equivalent to

$$a^2(a-b)(a-c) + b^2(b-c)(b-a) + c^2(c-a)(c-b) \geq 0,$$

which is the case $r = 2$ of Schur's inequality.

2.10 Challenges

2.1 A well-known express delivery service restricts the size of packages it will accept. Packages cannot exceed 165 inches in length plus girth, i.e., length $+ 2 \times$ width $+ 2 \times$ height ≤ 165. Find the dimensions of an acceptable package with maximum volume.

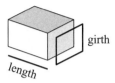

girth

length

Figure 2.19.

2.2 Use induction and the AM-GM inequality for two positive numbers to prove the AM-GM inequality for 2^k positive numbers, $k \geq 1$.

2.3 For nonnegative numbers x, y, z, prove that

(a) $x^3 + y^3 + z^3 + 3xyz \geq xy(x + y) + yz(y + z) + xz(x + z)$,

(b) $xyz \geq (x + y - z)(y + z - x)(z + x - y)$. [Hint: use Schur's inequality.]

2.4 Let K denote the area of a quadrilateral Q with sides lengths a, b, c, and d, in that order (i.e., a and c are opposite, as are b and d). Establish the following inequalities:

(a) $K \leq (ab + cd)/2$ and $K \leq (ad + bc)/2$,

(b) $K \leq (a + c)(b + d)/4$,

(c) $K \leq (a + b + c + d)^2/16$,

(d) $K \leq (a^2 + b^2 + c^2 + d^2)/4$.

We have equality in one part of (a) if and only if two opposite angles of Q are right angles, in (b) if and only if Q is a rectangle, and in (c) and (d) if and only if Q is a square. An ancient Egyptian approximation to the area of a quadrilateral is $K \approx (a + c)(b + d)/4$.

2.5 Prove that (a) if two fractions have the same denominator, then their mediant is the arithmetic mean and (b) if two fractions have the same numerator, then their mediant is the harmonic mean.

2.6 Prove the following generalization of the mediant property: if $a_1, a_2, \ldots, a_n$ are real numbers, and $b_1, b_2, \ldots, b_n$ are positive real numbers, then

$$\min_{1 \leq i \leq n} \frac{a_i}{b_i} \leq \frac{a_1 + a_2 + \cdots + a_n}{b_1 + b_2 + \cdots + b_n} \leq \max_{1 \leq i \leq n} \frac{a_i}{b_i}.$$

2.7 Let x, y be in $[0, 1]$. Create a visual proof that $\sqrt{xy} + \sqrt{(1-x)(1-y)}$ ≤ 1.

2.8 (a) Let a_1, a_2, b_1, b_2 be nonnegative real numbers. Prove that

$$\sqrt{a_1 a_2} + \sqrt{b_1 b_2} \leq \sqrt{(a_1 + b_1)(a_2 + b_2)}.$$

(b) Let $a_1, a_2, \ldots, a_n$ and $b_1, b_2, \ldots, b_n$ be nonnegative real numbers. Show that

$$(a_1 a_2 \cdots a_n)^{1/n} + (b_1 b_2 \cdots b_n)^{1/n}$$
$$\leq ((a_1 + b_1)(a_2 + b_2) \cdots (a_n + b_n))^{1/n}$$

so the sum of geometric means is less than or equal to the geometric mean of the sums.

2.9 Let a, b, c be positive numbers. Show that

$$\frac{1}{a+b} + \frac{1}{b+c} + \frac{1}{c+a} \geq \frac{9}{2(a+b+c)}.$$

2.10 Prove that for any integer $n > 1$, $n! < [(n+1)/2]^n$.

2.11 Prove that of all rectangular boxes with a given volume, the cube has minimum surface area.

2.12 Which is larger, $\sqrt{10} + \sqrt{18}$ or $\sqrt{12} + \sqrt{15}$?

2.13 Let $a \geq b \geq c \geq 0$ and $a + b + c \leq 1$. Give a visual proof that $a^2 + 3b^2 + 5c^2 \leq 1$. (Hint: begin with a square of side length $a + b + c$ [Jiang, 2007].)

2.14 Show that Schur's inequality implies Padoa's inequality.

2.15 Use the Chebyshev inequality to establish Mengoli's inequality (see Application 1.4).

2.16 Create a visual proof that for any $x \geq y \geq 0$, we have

$$\sqrt{3y^2 + x^2} \leq x + y \leq \sqrt{3x^2 + y^2}.$$

2.17 Matthias Roriczer, in his *Geometria Deutsch* (1486), presented the method illustrated in Figure 2.20 for constructing a square equal in area to a given equilateral triangle:

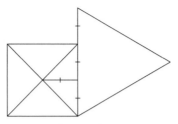

Figure 2.20.

Does the method work?

2.18 Let a, b, c be the sides of a triangle with $a \geq b \geq c$. Show that $a^2 + b^2 + c^2 \leq (a + b)(b + c)$.

2.19 If a, b, c are the sides of a right triangle with hypotenuse c, show that $ac + b^2 \leq 5c^2/4$.

2.20 The following pair of pictures illustrates a basic inequality [Kobayashi, 2002]. Which one? (The arrows indicate you are to fold on the dotted lines.)

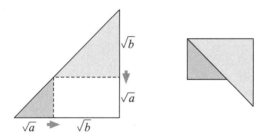

Figure 2.21.

2.21 (a) Let a, b, x be positive numbers, with $a < b$. Create a visual proof that
$$\frac{a}{a + x} < \frac{b}{b + x}.$$
       (b) Prove that if a, b, c are the sides of a triangle, so are $a/(a + 1)$, $b/(b + 1)$, and $c/(c + 1)$ [Martín and Plaza, 2008].

An open problem

Find and prove a four-variable version of Schur's inequality.

CHAPTER 3

Inequalities and the existence of triangles

Since the time of Euclid, geometers have studied procedures for constructing triangles given elements such as the three sides, two sides and an angle, and so on. The construction procedure usually has a constraint, such as using only an unmarked straightedge and compass to draw the triangle. There are constraints on the given elements as well, usually given as inequalities. For example, as noted in Section 1.1, a triangle with sides of length a, b, and c can be constructed if and only if the three triangle inequalities $a < b + c$, $b < c + a$, and $c < a + b$ hold.

In this chapter we examine inequalities among other elements of triangles that are necessary and sufficient for the existence of a triangle. We discuss the altitudes h_a, h_b, h_c; the medians m_a, m_b, m_c; and the angle-bisectors w_a, w_b, w_c.

Inequalities in the *Elements* of Euclid

In the first book of the *Elements*, Euclid states sixteen properties of equality and inequality. Of his five Common Notions, just one, the fifth, deals with inequality: *The whole is greater than the part.* Other properties of inequality appear in proofs of propositions. For example, in the proof of Proposition I.17 (In any triangle, the sum of any two angles is less than two right angles), Euclid uses the property that if $x < y$, then $x + z < y + z$ [Joyce].

3.1   Inequalities and the altitudes of a triangle

A triangle and its altitudes

Conditions on the altitudes h_a, h_b, and h_c in a triangle ABC with sides a, b, and c can be found by considering the area K of the triangle. Since

$$K = \frac{1}{2}ah_a = \frac{1}{2}bh_b = \frac{1}{2}ch_c,$$

we have

$$\frac{1}{h_a} = \frac{1}{2K}a, \qquad \frac{1}{h_b} = \frac{1}{2K}b, \qquad \frac{1}{h_c} = \frac{1}{2K}c,$$

so a triangle with sides $1/h_a$, $1/h_b$, $1/h_c$ must exist and be similar to the original triangle, with $1/2K$ the factor of similarity.

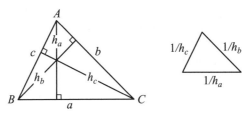

Figure 3.1.

So we must have $1/h_a < 1/h_b + 1/h_c$, $1/h_b < 1/h_c + 1/h_a$, and $1/h_c < 1/h_a + 1/h_b$ and (3.1) implies that if $a \leq b \leq c$, then $h_a \geq h_b \geq h_c$.

Existence of a triangle given a, b, and h_a

Clearly $b \geq h_a > 0$ is necessary. To show that it is also sufficient, we give a construction. Draw BC with length $|BC| = a$, and CD perpendicular to BC with $|CD| = h_a$, as shown in Figure 3.2. Draw a line parallel to BC at D, and locate vertex A at a distance of b from C. Thus $b \geq h_a > 0$ is necessary and sufficient for the existence of a triangle given a, b, and h_a.

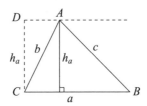

Figure 3.2.

Existence of a triangle given a, h_b, and h_c

Draw $|BC| = a$ and a semicircle with BC as its diameter. Then locate the feet of the altitudes h_b and h_c on the semicircle, and draw lines from B and C through those points to locate vertex A. (If one altitude, say h_b, equals a, then draw AC perpendicular to BC at C.) See Figure 3.3. Consequently, a sufficient (and necessary) condition for the existence of the triangle is $h_b \leq a$ and $h_c \leq a$, with at least one inequality strict. If the altitudes h_b and h_c intersect outside the semicircle, then triangle ABC is obtuse.

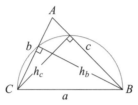

Figure 3.3.

More inequalities for the three altitudes

We now find upper bounds for the sum $h_a + h_b + h_c$ and the product $h_a h_b h_c$ of the three altitudes of a triangle with sides a, b, c in terms of the semiperimeter s and the area K [Bottema et al., 1968; Santaló, 2004].

Lemma 3.1. $h_a \leq \sqrt{s(s-a)}$, $h_b \leq \sqrt{s(s-b)}$, and $h_c \leq \sqrt{s(s-c)}$.

Proof. Using Heron's formula for the area K of the triangle (for a proof of Heron's formula, see Application 4.1), we have

$$h_a = \frac{2K}{a} = \frac{2}{a}\sqrt{s(s-a)(s-b)(s-c)} = \sqrt{s(s-a)}\frac{\sqrt{(s-b)(s-c)}}{a/2}$$

The AM-GM inequality applied to $s - b$ and $s - c$ yields

$$\sqrt{(s-b)(s-c)} \leq \frac{s-b+s-c}{2} = \frac{a}{2},$$

so $h_a \leq \sqrt{s(s-a)}$. The lemma gives

Theorem 3.1. $h_a + h_b + h_c \leq \sqrt{3}s$ and $h_a h_b h_c \leq sK$.

Proof. Using the arithmetic mean-root mean square inequality for the three numbers $\sqrt{s(s-a)}$, $\sqrt{s(s-b)}$, and $\sqrt{s(s-c)}$ (see Section 2.6), we have

$$h_a + h_b + h_c \leq \sqrt{s(s-a)} + \sqrt{s(s-b)} + \sqrt{s(s-c)}$$

$$\leq 3\sqrt{\frac{s(s-a) + s(s-b) + s(s-c)}{3}} = \sqrt{3s},$$

and

$$h_a h_b h_c \leq \sqrt{s(s-a)}\sqrt{s(s-b)}\sqrt{s(s-c)} = sK.$$

Another proof of Theorem 3.1 can be obtained by using Theorem 3.3 (in Section 3.3) with the observation that the length of each altitude is less than or equal to the length of the corresponding angle-bisector.

Altitudes, sides and angles

If $a \leq b \leq c$ are the sides of a triangle, we can now show that $\cos A \geq \cos B \geq \cos C$. From Figure 3.1, we see $a = b \cos C + c \cos B$, $b = a \cos C + c \cos A$, and $c = a \cos B + b \cos A$. Hence

$$0 \leq (b-a)\cos C \leq c(\cos A - \cos B), \text{ so } \cos B \leq \cos A$$

and

$$0 \leq (c-b)\cos A \leq a(\cos B - \cos C), \text{ so } \cos C \leq \cos B.$$

3.2   Inequalities and the medians of a triangle

In the triangle ABC, the *medians* are the segments joining each vertex to the midpoint of the opposite side, and have lengths m_a, m_b, and m_c. The three medians intersect at a point, the *centroid* or center of gravity of the triangle, and the distance along a median from a vertex to the centroid is two-thirds the length of that median. See Figure 3.4.

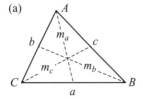
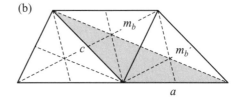

Figure 3.4.

Using the law of cosines, it is easy to show that the length of a median is a function of the sides a, b, and c, i.e., $m_a^2 = (2b^2 + 2c^2 - a^2)/4$, and similarly for m_b^2 and m_c^2. Consequently $m_a^2 - m_b^2 = 3(b^2 - a^2)/4$, so if the sides satisfy $a \le b \le c$, then $m_a \ge m_b \ge m_c$. Also, $m_a^2 + m_b^2 + m_c^2 = 3(a^2 + b^2 + c^2)/4$.

Theorem 3.2. $3s/2 \le m_a + m_b + m_c \le 2s$.

Proof. After arranging three copies of a triangle into a trapezoid in Figure 3.4, we apply the triangle inequality to the shaded triangle to conclude $2m_b \le a + c$. Similarly, $2m_a \le b + c$ and $2m_c \le a + b$, so

$$m_a + m_b + m_c \le a + b + c = 2s. \tag{3.1}$$

To find the lower bound on $m_a + m_b + m_c$, we arrange six copies of the triangle into the hexagon in Figure 3.5, and note that the shaded triangle has sides $2m_a$, $2m_b$, and $2m_c$, and medians $3a/2$, $3b/2$, and $3c/2$. Applying (3.2) to this triangle yields $3(a + b + c)/2 \le 2(m_a + m_b + m_c)$, which completes the proof.

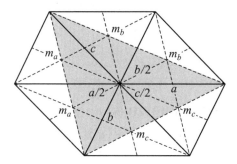

Figure 3.5.

In Application 2.4 we saw that the area K of a triangle is related to its semiperimeter s by the inequality $K \le \sqrt{3}s^2/9$. If K denotes the area of triangle ABC in Figure 3.4, then the area of the shaded triangle in Figure 3.5 is $3K$, its semiperimeter is $m_a + m_b + m_c$, and hence

$$K \le \frac{\sqrt{3}}{27}(m_a + m_b + m_c)^2.$$

Also, the area of a triangle with sides m_a, m_b, and m_c (called the *median triangle* of ABC) is $3K/4$.

Existence of a triangle given m_a, m_b, and m_c

If we can construct a triangle with sides $2m_a$, $2m_b$, and $2m_c$, then we can construct the shaded triangle in Figure 3.5 and find its centroid. Then a, b, c are the lengths of the segments between the centroid and the vertices. Hence we need the inequalities

$$m_a < m_b + m_c, \qquad m_b < m_c + m_a, \qquad \text{and} \qquad m_c < m_a + m_b.$$

Existence of a triangle given a, b, and m_a

If we can construct a triangle with sides $a/2$, b, and m_a, then we can construct the gray triangle in Figure 3.6, and double its base to obtain the desired triangle. Hence we need the three inequalities

$$m_a < b + a/2, \qquad a/2 < b + m_a \qquad \text{and} \qquad b < m_a + a/2.$$

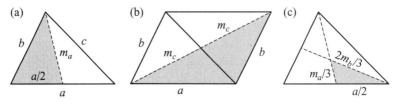

Figure 3.6.

Existence of a triangle given a, b, and m_c

If we can construct a triangle with sides a, b, and $2m_c$, we can construct the gray triangle in Figure 3.6b, complete the parallelogram and draw its other diagonal to obtain the desired triangle. Hence we need the three inequalities

$$2m_c < a + b, \qquad a < b + 2m_c, \qquad \text{and} \qquad b < a + 2m_c.$$

Existence of a triangle given a, m_a, and m_b

As before, if we can construct a triangle with sides $a/2$, $m_a/3$, and $2m_b/3$, we can construct the gray triangle in Figure 3.6c, and extend the sides of length $a/2$ and $m_a/3$ to locate the other two vertices of the triangle. Hence we need the three inequalities

$$\frac{a}{2} < \frac{m_a}{3} + \frac{2m_b}{3}, \qquad \frac{m_a}{3} < \frac{a}{2} + \frac{2m_b}{3}, \qquad \text{and} \qquad \frac{2m_b}{3} < \frac{a}{2} + \frac{m_a}{3}.$$

3.3 Inequalities and the angle-bisectors of a triangle

Let w_a, w_b, and w_c denote the lengths of the line segments that bisect the angles of triangle ABC and terminate on the opposite sides. We now find bounds on $w_a + w_b + w_c$ and $w_a w_b w_c$ in terms of the semiperimeter s and the area K.

Theorem 3.3. $w_a + w_b + w_c \leq \sqrt{3}s$ and $w_a w_b w_c \leq sK$.

Proof. We begin by finding explicit expressions for w_a, w_b, and w_c in terms of a, b, and c. In triangle ABC (see Figure 3.7), we first extend BC a length b to the point F, and draw AF. Let CD with length w_c bisect $\angle C$, extend CD to intersect the circumcircle at E, and draw BE and AE. Because triangle ACF is isosceles with $\angle AFC$ one-half of $\angle C$, AF is parallel to CE.

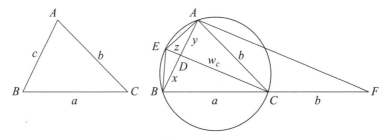

Figure 3.7.

Since triangle DBC is similar to triangle ABF, we have $x/c = a/(a+b)$, so $x = ac/(a+b)$ and $y = c - x = bc/(a+b)$. Since triangle DAC is similar to triangle BEC because $\angle DAC = \angle BEC$ and $\angle ACD = \angle ECB$, we have $(z + w_c)/b = a/w_c$, and hence $w_c^2 + zw_c = ab$. But the intersecting chords theorem says that when two chords intersect inside a circle, the products of their segments are equal. Applied to AB and CE this yields $zw_c = xy = abc^2/(a+b)^2$, and hence

$$w_c = \frac{2\sqrt{ab}}{a+b}\sqrt{s(s-c)}.$$

But $2\sqrt{ab} \leq a+b$ by the AM-GM inequality, so $w_c \leq \sqrt{s(s-c)}$. Similarly $w_a \leq \sqrt{s(s-a)}$ and $w_b \leq \sqrt{s(s-b)}$, so by the arithmetic mean-root mean square inequality we have

$$w_a + w_b + w_c \leq \sqrt{s(s-a)} + \sqrt{s(s-b)} + \sqrt{s(s-c)}$$

$$\leq 3\sqrt{\frac{s(s-a) + s(s-b) + s(s-c)}{3}} = \sqrt{3s}$$

and

$$w_a w_b w_c \leq \sqrt{s(s-a)}\sqrt{s(s-b)}\sqrt{s(s-c)} = sK.$$

The bounds in Theorem 3.3 are the same as the ones for the altitudes in Theorem 3.1.

Existence of a triangle given a, h_a, and w_a

From Figure 3.8 the only inequality required is $w_a \geq h_a$.

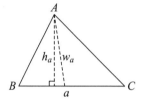

Figure 3.8.

Unlike earlier existence examples, we do not have a construction proce-dure in this case. The same figure provides the inequality required for the existence of a triangle given a, h_a, and m_a is $m_a \geq h_a$.

Existence of a triangle given a, h_b, and w_c

Draw $|BC| = a$ and a semicircle with BC as its diameter. Then locate D at a distance h_b from B as shown in Figure 3.9. Bisect $\angle BCD$ and locate E at a distance w_c from C on the angle bisector. Draw CD and BE and extend them to meet at A. This construction is possible if and only if

$$h_a \leq a \quad \text{and} \quad w_c < \left[h_b^2 + \left(a + \sqrt{a^2 - h_b^2} \right)^2 \right]^{1/2}.$$

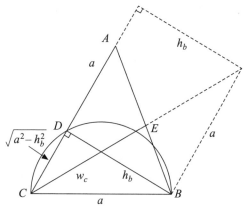

Figure 3.9.

Ordering of sides and angle-bisectors

Earlier in this chapter we showed that if $a \leq b \leq c$ then $h_a \geq h_b \geq h_c$ and $m_a \geq m_b \geq m_c$. We now show that the same inequalities hold for the angle-bisectors: $w_a \geq w_b \geq w_c$.

If $0 < a \leq b \leq c$, then $c - a \geq c - b$ and $2\sqrt{ac} \leq 2\sqrt{bc}$, and hence $(c - a)/(2\sqrt{ac}) \geq (c - b)(2\sqrt{bc})$. Let $\alpha = \arctan(c - a)/(2\sqrt{ac})$ and $\beta = \arctan(c - b)/(2\sqrt{bc})$, as illustrated in Figure 3.10. Since $\beta \leq \alpha$, it follows that $\cos \beta = 2\sqrt{bc}/(b + c) \geq 2\sqrt{ac}/(a + c) = \cos \alpha$.

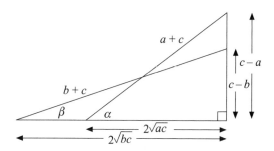

Figure 3.10.

But $a \leq b$ implies $\sqrt{s(s - a)} \geq \sqrt{s(s - b)}$, and thus

$$w_a = \frac{2\sqrt{bc}}{b + c}\sqrt{s(s - a)} \geq \frac{2\sqrt{ac}}{a + c}\sqrt{s(s - b)} = w_b.$$

The proof that $w_b \geq w_c$ is similar. The proof can be modified to show that $0 < a < b$ implies $w_a > w_b$.

3.4   The Steiner-Lehmus theorem

It is an easy exercise to prove that, in an isosceles triangle, the angle-bisectors of the equal angles are equal. The converse is not so easy to prove, and has become known as the *Steiner-Lehmus theorem*. In 1840, C. L. Lehmus asked the Swiss geometer Jakob Steiner, whom we shall meet again in Chapter 5, for a purely geometrical proof, as algebraic proofs are relatively simple once one knows the lengths of the angle bisectors in terms of the sides (see Challenge 3.11). H. S. M. Coxeter and S. L. Greitzer [Coxeter and Greitzer, 1967] write that this result is one that seems "to exert a peculiar fascination on anybody who happens to stumble on" it. Since 1840, hundreds of different proofs have been published, and they continue to appear with some regularity. Our proof is based on the result in the previous section concerning the order of the lengths of the angle bisectors.

The Steiner-Lehmus Theorem 3.4. *If two angle bisectors of a triangle are equal, then the triangle is isosceles.*

Proof. We need to show that $w_a = w_b$ implies $a = b$. If $a < b$, then $w_a > w_b$, and if $a > b$, then $w_a < w_b$. Thus by *reductio ad absurdum*, $w_a = w_b$ implies $a = b$.

Given the simplicity of this proof, one might wonder about Coxeter and Greitzer's claim about the "peculiar fascination" with this theorem for some mathematicians. But Lehmus requested a *purely geometric* proof—ours is not because we used algebra to compute the lengths of the angle bisectors. In addition, the search has been for direct geometric proofs—nearly all known geometric proofs are like ours indirect, *reductio ad absurdum,* proofs.

Many more inequalities relating the sides, angles, altitudes, medians, and angle bisectors of triangles can be found in [Bottema et al, 1968] and [Mitrinovic et al, 1989].

3.5   Challenges

Note: In these Challenges, a, b, c denote the sides, h_a, h_b, h_c the altitudes, m_a, m_b, m_c the medians, w_a, w_b, w_c, the angle-bisectors, r the radius of the incircle, and R the radius of the circumcircle of triangle ABC.

3.1 If ABC is an acute triangle with A, B, and C given in radians, show that there exists a triangle whose sides have lengths A, B, and C.

3.2 When do A, h_a, and r form a triangle?

3.3 Using a picture, find necessary and sufficient conditions for the existence of a triangle given b, c, and h_a. What happens if a, h_a, and h_b are given?

3.4 Given a, m_b, and m_c, when does a unique triangle exist?

3.5 Given a, b, and R, when is a triangle determined?

3.6 What are the inequalities among a, b, m_b, and R in a triangle ABC?

3.7 Show that necessary and sufficient conditions for the existence of a triangle given a, $b + c$, and the angle A are $a < b + c$ and $\sin(A/2) \leq a/(b + c)$.

3.8 Show that if ABC is a triangle, then $\sin A$, $\sin B$, and $\sin C$ are the lengths of the sides of another triangle. Do $\cos A$, $\cos B$, and $\cos C$ form a triangle?

3.9 When do a, A, and h_a form a triangle?

3.10 If a, b, c and h_a, h_b, and h_c are in a triangle, show that $a^2 h_a, b^2 h_b$, and $c^2 h_c$ form a triangle.

3.11 Give a direct algebraic proof of the Steiner-Lehmus theorem by showing that $w_a = w_b$ implies $a = b$.

CHAPTER 4

Using incircles and circumcircles

For many geometric inequalities, the strategy of inscribing or circumscribing a figure can be useful (recall, for example, Application 1.1, Sections 1.3 and 1.6, and Theorem 2.3). Of all such inscribed or circumscribed figures, the circle plays a central role, and results in a variety of inequalities relating the radius of the circle to numbers associated with the given figure, such as side lengths, perimeter, area, etc.

The triangle is exceptional because every triangle possesses a circle passing through the vertices of the triangle, the *circumcircle*, whose center is the *circumcenter* of the triangle, and a circle inside the triangle and tangent to its three sides, the *incircle*, whose center is the *incenter* of the triangle.

> **Inscribing and circumscribing in the *Elements* of Euclid**
>
> Among the thirteen books of the *Elements* of Euclid (circa 300 BCE), there is one, Book IV, devoted to inscribing and circumscribing figures. This book starts with two basic definitions [Joyce]:
>
> **Definition 1**. A rectilinear figure is said to be *inscribed in a rectilinear figure* when the respective angles of the inscribed figure lie on the respective sides of that in which it is inscribed.
>
> **Definition 2**. Similarly a figure is said to be *circumscribed about a figure* when the respective sides of the circumscribed figure pass through the respective angles of that about which it is circumscribed.
>
> After five complementary definitions, Euclid proves sixteen propositions. Euclid mainly considers problems of inscribing or circumscribing circles about triangles, squares, regular pentagons and hexagons, and regular 15-gons.

Only special n-gons with $n \geq 4$ possess a circumcircle or an incircle. Those that do often give nice inequalities. Moreover, inscribing or circumscribing a polygon in or about a circle is an ancient and classical method for approximating the circumference and area of a circle, as we saw in Chapter 1.

If inequalities are impossible, equality follows

A celebrated argument in Archimedes' *On the Measurement of the Circle* is the proof of Proposition 1 that "the area [A] of any circle is equal to [the area K of] a right-angled triangle in which one of the sides about the right angle is equal to the radius [r], and the other to the circumference [C], of the circle" [Heath, 1953]. Archimedes proceeds to show that the two possibilities $A > K$ and $A < K$ yield contradictions. Supposing that $A > K$, Archimedes knew that by inscribing regular n-gons one could find an n-gon of area Q_n such that $A > Q_n > K$. But if P_n denotes the perimeter and a the inradius of the n-gon, then $Q_n = P_n a/2 < Cr/2 = K$, a contradiction. Archimedes disposed of the inequality $A < K$ analogously.

4.1   Euler's triangle inequality

Euler's theorem for the triangle states that the distance d between the circumcenter and the incenter is given by $d^2 = R(R - 2r)$, where R and r are the circumradius and inradius, respectively. An immediate consequence of this theorem is $R \geq 2r$, which is often referred to as *Euler's triangle inequality*. Coxeter [Coxeter, 1985] notes that although Euler published this inequality in 1767 [Euler, 1767], it had appeared in 1746 in a publication by William Chapple. Here we will use visual proofs to establish three lemmas that reduce the proof of Euler's triangle inequality to simple algebra. The proof is based on one that appears in [Klamkin, 1967].

Let a, b, c denote the lengths of the sides of the triangle, and employ the Ravi substitution from Section 1.6: $a = x + y, b = y + z, c = z + x$, as illustrated in Figure 4.1. Then, as we saw in Application 1.5, Padoa's inequality holds, which can be expressed compactly as $abc \geq 8xyz$. We now show that this inequality is equivalent to Euler's triangle inequality $R \geq 2r$.

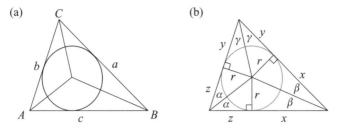

Figure 4.1.

The three lemmas we prove are of interest in their own right—for example, in Application 4.1 we use two of them to establish Heron's formula for the area of a triangle. The proofs of the lemmas are elementary, using nothing more sophisticated than similarity of triangles. The first expresses the area K of the triangle in terms of the three side lengths a, b, c and the circumradius R. The second, whose proof uses a rectangle composed of triangles similar to the right triangles in Figure 4.1b, expresses the product xyz in terms of the inradius r and the semiperimeter $s = x + y + z$. The third gives the area K in terms of r and s.

Lemma 4.1. $4KR = abc$.

Proof.

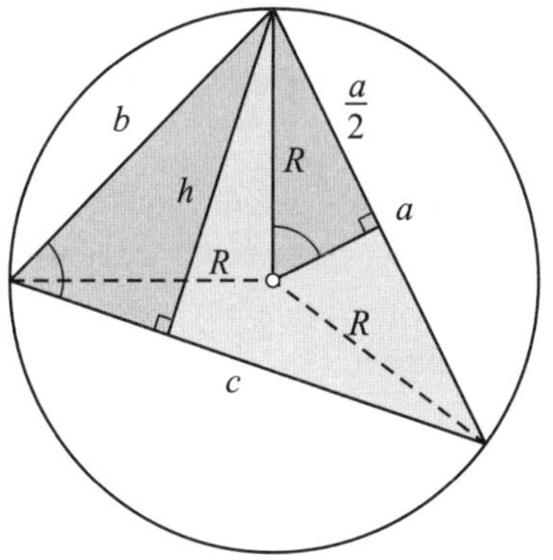

$$\frac{h}{b} = \frac{a/2}{R} \quad \Rightarrow \quad h = \frac{1}{2}\frac{ab}{R}$$

$$K = \frac{1}{2}hc = \frac{1}{4}\frac{abc}{R}$$

Figure 4.2.

Lemma 4.2. $xyz = r^2(x + y + z) = r^2 s$.

Proof. Letting w denote $\sqrt{r^2 + z^2}$, we have

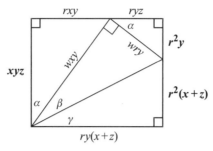

Figure 4.3.

Lemma 4.3. $K = r(x + y + z) = rs.$

Proof.

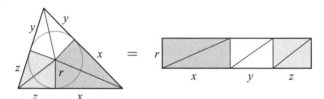

Figure 4.4.

We now show that Padoa's inequality $abc \geq 8xyz$ is easily seen to be equivalent to Euler's triangle inequality.

Theorem 4.1. *If R and r denote the circumradius and inradius, respectively, of a triangle, then $R \geq 2r$.*

Proof. From Padoa's inequality and Lemma 4.1, we obtain $4KR \geq 8xyz$. Lemma 4.2 then gives $4KR \geq 8r^2s$ and with Lemma 4.3 we have $4KR \geq 8Kr$, from which $R \geq 2r$ follows. The steps can be reversed to show that Euler's inequality implies Padoa's inequality.

Euler's triangle inequality cannot be improved for general triangles, since $R = 2r$ when the triangle is equilateral. However, for right triangles, we have $R \geq (1 + \sqrt{2})r$, with equality for isosceles right triangles. We leave the proof of this inequality as Challenge 4.9.

Application 4.1.   *Heron's formula*

Lemmas 4.2 and 4.3 together yield a simple proof of Heron's remarkable formula for the area K of the triangle with side lengths a, b, c:

$$K = \sqrt{s(s - a)(s - b)(s - c)},$$

where s denotes the semiperimeter, $s = (a + b + c)/2 = x + y + z$ [Nelsen, 2001]. Since $s - a = z$, $s - b = x$, and $s - c = y$, Lemma 4.2 yields $r^2 s = (s - a)(s - b)(s - c)$, or $(rs)^2 = s(s - a)(s - b)(s - c)$. But from Lemma 4.3 we have $K = rs$, from which Heron's formula for K now follows.

We now use Heron's formula and Lemmas 4.2 and 4.3 to prove the following companion to Theorem 2.3.

Theorem 4.2. *Among all triangles that can be circumscribed about a given circle, the equilateral triangle has the least area.*

Proof. Label the parts of the triangle as illustrated in Figure 4.1, and let K denote its area. From Lemma 4.2, we have $xyz = r^2(x + y + z)$, or

$$\frac{1}{r^2} = \frac{x + y + z}{xyz} = \frac{1}{yz} + \frac{1}{xz} + \frac{1}{xy}.$$

From Lemma 4.3 and Heron's formula, we have $K^2 r^2 = K^4/s^2 = (xyz)^2$, and thus the AM-GM inequality yields

$$\frac{1}{K^2 r^2} = \frac{1}{yz} \cdot \frac{1}{xz} \cdot \frac{1}{xy} \le \left(\frac{(1/xy) + (1/yz) + (1/xz)}{3} \right)^3 = \frac{1}{27r^6}.$$

Consequently K is a minimum when $1/K^2 r^2$ is a maximum, which occurs when we have equality. Hence the area of the triangle is smallest when $x = y = z$, or when the triangle is equilateral.

Inscribing and circumscribing gone wrong

In [Aczél, 1980], János Aczél relates the story of a person who found the area of a circle of radius 1 by inscribing and circumscribing squares with areas 2 and 4 (as shown below) and concluding that the area of the circle must be the arithmetic mean of 2 and 4, namely 3.

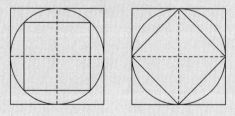

Corollary 4.1. *If K and r denote the area and inradius, respectively, of a triangle, then $K \ge 3\sqrt{3}r^2$.*

Corollaries 2.1 and 4.1 combine to yield the next corollary, which strengthens Euler's triangle inequality.

Corollary 4.2. *If K, r, and R denote the area, inradius, and circumradius, respectively, of a triangle, then* $R \geq 2\sqrt{K}/3^{3/4} \geq 2r$.

4.2   The isoperimetric inequality

This result, the queen of geometric inequalities, is called *isoperimetric* because it says that among all closed curves with a given perimeter L, the one that encloses maximum area is the circle. Specifically, given a region in the plane with perimeter L and area K, what can we say in general about the relationship between L and K? For a circle of radius r, $L = 2\pi r$ and $K = \pi r^2$, so $L^2 = 4\pi K$. The isoperimetric theorem establishes that, in general, we have inequality between L^2 and $4\pi K$.

The Isoperimetric Theorem 4.3. Given a region in the plane with perimeter L and area K,

$$L^2 \geq 4\pi K,$$

with equality in the case of the circle.

Proof. Closed curves may be uniformly approximated by polygons, i.e., polygons whose perimeters and areas tend, respectively, to the perimeter and area of the figure. Thus it suffices to prove the inequality for arbitrary n-gons. The following elegant proof is from [Dergiades, 2002], based on one which appears in [Bonnesen, 1929] and [Bonnesen and Fenchel, 1948]. In Figure 4.5, we have a method for a bisection of a circular disk.

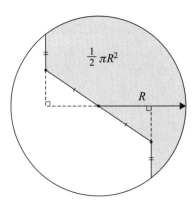

Figure 4.5.

Since it suffices to consider only convex n-gons (see Example 5.3 in the next chapter), we begin with a convex n-gon with perimeter L and area K, as illustrated in Figure 4.6, and follow this sequence of steps:

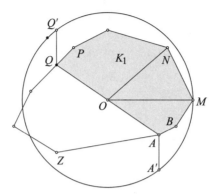

Figure 4.6.

1. Choose a vertex of the n-gon, label it A, locate a point Q on the polygon such that the distance along the edges of the n-gon from A to Q is $L/2$, and join A to Q with a line segment. This segment splits the n-gon into two smaller polygons, one of which has area $K_1 \geq K/2$. Denote this polygon $AB \cdots PQA$.

2. Let O be the midpoint of segment AQ, let M be the vertex of $AB \cdots PQA$ farthest from O, set $R = |OM|$, and draw a circle with center O and radius R. From points A and Q draw segments AA' and QQ' perpendicular to OM meeting the circle at points A' and Q', respectively. The portion of the circle passing through the points A, A', M, Q', Q, and O has area $\pi R^2 / 2$.

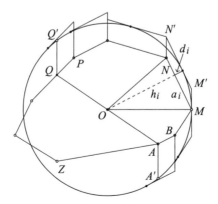

Figure 4.7.

3. Now construct parallelograms outwardly on the sides of $AB\cdots PQA$ tangent to the circle, as illustrated in Figure 4.7, with bases segments such as MN and sides MM' and NN' parallel to AA'. Let $a_i = |MN|$ let h_i denote the altitude of triangle OMN and d_i the height of the parallelogram $MM'NN'$ (so $h_i + d_i = R$), and similarly for other triangles with vertices O and pairs of adjacent vertices of the n-gon.

Since K_1 is the sum of the area of the triangles $OAB, \ldots, OMN, \ldots,$ OPQ, we have

$$K_1 = \frac{1}{2}\sum_i a_i h_i,$$

and if K_2 denotes the sum of the area of the parallelograms, then

$$K_2 = \sum_i a_i d_i = \sum_i a_i (R - h_i) = R \cdot \frac{L}{2} - 2K_1.$$

Since $K_1 + K_2 \geq \pi R^2/2$, we have $RL/2 - K_1 \geq \pi R^2/2$, so $\pi R^2 - LR + 2K_1 \leq 0$. Completing the square in the quadratic yields

$$\pi\left(R - \frac{L}{2\pi}\right)^2 - \left(\frac{L^2}{4\pi} - 2K_1\right) \leq 0,$$

so $L^2 \geq 4\pi \cdot 2K_1 \geq 4\pi K$, as required.

The historical background of the isoperimetric theorem

Today, the isoperimetric theorem states that *of all plane figures with a given perimeter, the circle has the greatest area, and of all plane figures with a given area, the circle has the least perimeter*. But this result has behind it more than 2000 years of mathematical work. Euclid knew the isoperimetric theorem for rectangles; Archimedes considered isoperimetric problems for n-gons; Zenodorus (circa 200-140 BCE) wrote a book, now lost, *On Isometric Figures*, and his results were quoted by Pappus of Alexandria (circa 280-350). Years later Simon Lihuilien and Jakob Steiner made contributions to this topic, and even Steiner's error with the isoperimetric theorem motivated other mathematicians such as Karl Weierstrass to introduce new analytical tools, such as the calculus of variations.

The isoperimetric theorem and the planimeter

A planimeter is a mechanical instrument used to compute an approximation to the area of a plane region bounded by a simple closed curve

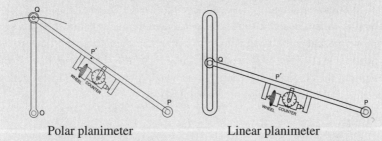

Polar planimeter   Linear planimeter

In the linear case, we have an arm of length l with a rolling wheel attached to an interior point of the arm, so as the end of the arm traces the curve, the amount σ the wheel rolls is proportional to the area K of the region. However, $\sigma \leq L$ (the perimeter) since the wheel records only the component of motion perpendicular to the arm. Then one obtains $K = l\sigma - \pi l^2$, so, completing the square, $\sigma^2 - 4\pi K = (K - \pi l^2)/l^2$, and hence $L^2 - 4\pi K \geq (K - \pi l^2)/l^2 \geq 0$. It is interesting that a mechanical device can experimentally confirm the isoperimetic inequality. For more on how planimeters work, see [Leise, 2007; Foote].

Isoperimetric results in space

If C denotes a three-dimensional convex body with volume V and surface area S, let D denote the *diameter* of C, the maximum distance between two points of C, and E the *width* of C, the minimum distance between two points of C such that C lies between two parallel planes, each containing one of the two points. Then

$$V \leq S^{3/2}/6\sqrt{\pi}, \quad V \leq \pi D^3/6, \quad V \leq S^2/6\pi E,$$

and

$$S \leq \pi D^2, \quad V \leq \sqrt{A_{xy} A_{yz} A_{xz}},$$

where A_{xy}, A_{yz}, and A_{xz}, are the area of the projections of C with respect to the coordinate planes. When C is a sphere, all of the above inequalities are equalities (see [Mitrinović et al., 1989] for many inequalities of this type).

4.3   Cyclic, tangential, and bicentric quadrilaterals

A quadrilateral that possesses a circumcircle is called *cyclic*, a quadrilateral that possesses an incircle is called *tangential*, and a tangential cyclic quadrilateral is called *bicentric* (or a *cyclic-inscriptible* quadrilateral). In this section we present some inequalities for cyclic and bicentric quadrilaterals. In a quadrilateral Q, we will denote the lengths of the sides by a, b, c, d, the lengths of the diagonals by p and q, as shown in Figure 4.8a, and let K denote the area, R the circumradius, and r the inradius.

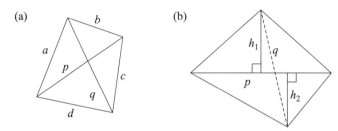

Figure 4.8.

Lemma 4.4. *If Q is cyclic, then $2R^2 \geq K$.*

Proof. The area K of the convex quadrilateral Q in Figure 4.8b is $K = p(h_1 + h_2)/2$, but $h_1 + h_2 \leq q$ and hence $K \leq pq/2$. Equality holds if and only if the diagonals of Q are perpendicular. But if Q is cyclic, then p and q are each less than or equal to the diameter $2R$, and hence $K \leq 2R^2$, as claimed.

Lemma 4.5. *If Q is bicentric, then $K \geq 4r^2$.*

Proof. In Figure 4.9, $2\theta + 2\phi = \pi$, so $\theta + \phi = \pi/2$ and hence the shaded triangles are similar. Thus $x/r = r/z$, or $r^2 = xz$. Similarly, $r^2 = ty$. Hence

$$K = r(x+y+z+t) = 2r \left(\frac{x+z}{2} + \frac{t+y}{2} \right) \geq 2r \left(\sqrt{xz} + \sqrt{ty} \right) = 4r^2.$$

We have equality if and only if $x = z = y = t = r$, i.e., when Q is a square.

The two lemmas now yield

Theorem 4.4. *If Q is bicentric, then $R^2 \geq K/2 \geq 2r^2$.*

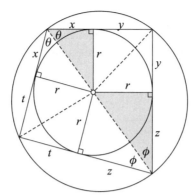

Figure 4.9.

Figure 4.9 also illustrates the useful property that for any tangential quadrilateral, the sums of the lengths of pairs of opposite sides are equal, i.e., $a + c = b + d$.

A classical result about cyclic quadrilaterals is *Ptolemy's theorem*: If Q is cyclic, then $pq = ac + bd$, proved in Section 6.1. Here is a companion result about the quotient of the lengths of the diagonals [Alsina and Nelsen, 2007a].

Lemma 4.6. *If Q is cyclic, then* $\dfrac{p}{q} = \dfrac{ad + bc}{ab + cd}$.

(a) 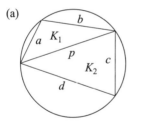   (b)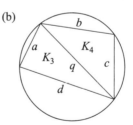

Figure 4.10.

Proof. Using the result of Lemma 4.1 to evaluate the areas of the triangles formed by sides and diagonals of Q (see Figure 4.10), we have [Alsina and Nelsen, 2007a; Weisstein]

$$K = K_1 + K_2 = \frac{pab}{4R} + \frac{pcd}{4R} = \frac{p(ab + cd)}{4R},$$

$$K = K_3 + K_4 = \frac{qad}{4R} + \frac{qbc}{4R} = \frac{q(ad + bc)}{4R},$$

and hence $p(ab + cd) = q(ad + bc)$, and the result follows.

Ptolemy's theorem and the last lemma combine to yield the following theorem stating the lengths of the diagonals of a cyclic quadrilateral Q in terms of the lengths of the sides, and a relationship similar to Lemma 4.1 for the area and circumradius of Q.

Theorem 4.5. *If Q is cyclic, then*

$$p = \sqrt{\frac{(ac + bd)(ad + bc)}{ab + cd}}, \quad q = \sqrt{\frac{(ac + bd)(ab + cd)}{ad + bc}},$$

and

$$4KR = \sqrt{(ab + cd)(ac + bd)(ad + bc)}.$$

Proof.

$$p^2 = pq \cdot \frac{p}{q} = \frac{(ac + bd)(ad + bc)}{ab + cd},$$

$$q^2 = pq \cdot \frac{q}{p} = \frac{(ac + bd)(ab + cd)}{ad + bc},$$

$$K^2 = \frac{pq(ab + cd)(ad + bc)}{(4R)^2} = \frac{(ac + bd)(ab + cd)(ad + bc)}{16R^2}.$$

As an application of Theorem 4.5, we have

Corollary 4.3. *If Q is cyclic, then $2K \le \sqrt[3]{(ab + cd)(ac + bd)(ad + bc)}$.*

Proof. From Lemma 4.4, we have

$$8K^2 = \frac{(ab + cd)(ac + bd)(ad + bc)}{2R^2} \le \frac{(ab + cd)(ac + bd)(ad + bc)}{K},$$

so $8K^3 \le (ab + cd)(ac + bd)(ad + bc)$, as required.

Using Ptolemy's theorem for cyclic quadrilaterals, the fact that $a + c = b + d$ for tangential quadrilaterals, and the AM-GM inequality, we can prove the following result [Klamkin, 1967]:

Corollary 4.4. *If Q is bicentric, then $(a + b + c + d)^2 \ge 8pq$.*

Proof. $8pq = 2(4ac + 4bd) \le 2[(a + c)^2 + (b + d)^2] = (a + b + c + d)^2$.

4.4 Some properties of *n*-gons

We begin with some results about triangles.

Theorem 4.6. *Among all triangles having a specified base and opposite vertex angle, the isosceles triangle has the greatest inradius.*

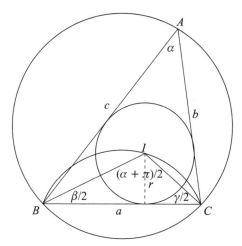

Figure 4.11.

Proof. Let *a* denote the length of the specified base *BC*, and α the measure of vertex angle *A*, as illustrated in Figure 4.11. If *I* denotes the incenter of *ABC*, then the measure of angle *BIC* is $\pi - (\beta/2 + \gamma/2) = (\alpha + \pi)/2$. Consequently the incenter *I* is located on a circle passing through *B* and *C*, and thus the inradius *r* is a maximum when triangle *BIC*, and hence triangle *ABC*, is isosceles.

The same figure provides a proof of the following corollary as well.

Corollary 4.5. *Among all triangles having a specified base and opposite vertex angle, the isosceles triangle has the greatest area.*

An analogous result holds for the perimeter of the triangle.

Theorem 4.7. *Among all triangles having a specified base and opposite vertex angle, the isosceles triangle has the greatest perimeter.*

Proof. Again, we let *a* denote the length of the specified base and α the measure of the vertex angle, as illustrated in Figure 4.11. To maximize the perimeter, we need only maximize $b + c$, or to maximize $(b + c)^2/2$.

The arithmetic mean-root mean square inequality (part of (1.1)), the law of cosines, and Lemma 4.1 yield

$$\frac{(b+c)^2}{2} \leq b^2 + c^2 = a^2 + 2bc\cos\alpha = a^2 + \frac{8KR\cos\alpha}{a}$$

with equality if and only if $b = c$. Hence the isosceles triangle has the greatest perimeter, as claimed.

Theorem 4.8. *A cyclic* n-*gon (an* n-*gon possessing a circumcircle) has larger area than any other* n-*gon with sides of the same length in the same order.*

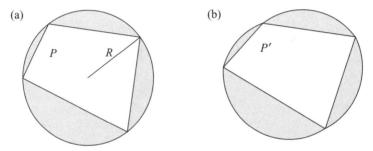

Figure 4.12.

Proof. Let P be an n-gon inscribed in a circle of radius R, as illustrated in Figure 4.12a. Let P' be an n-gon that is not congruent to P, but with sides of the same length and in the same order as P. On each side of P' construct circular segments (in gray) of radius R congruent to those surrounding P, as illustrated in Figure 4.12b, so both P and P' are surrounded by circular arcs with total length $2\pi R$. By the isoperimetric inequality (Section 4.2), the total area of P' and the gray segments is less than the total area, πR^2, of P and the gray segments. After subtracting the area of the gray segments, the area of P' must be less than the area of P.

As a result of Theorem 4.8, the hypothesis that Q is cyclic can be dropped from Corollary 4.3 and we can conclude that $2K \leq \sqrt[3]{(ab+cd)(ac+bd)(ad+bc)}$ holds for any quadrilateral Q.

The result in Theorem 4.8 is not true if the phrase "in the same order" is omitted. See Challenge 4.6.

4.5 Areas of parallel domains

Consider a figure F with area K bounded by a closed curve C of finite length L. For any real $r > 0$, denote by F_r the *parallel domain of F at a distance r*, i.e., the set of points whose distances from F are not greater than r. See Figure 4.13 for examples for (a) a convex and (b) a nonconvex polygon F (the striped polygon) and the corresponding F_r (the entire region):

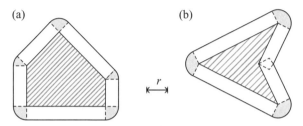

Figure 4.13.

Let K_r denote the area of F_r. Then $K_r \leq K + Lr + \pi r^2$, with equality if and only if C is convex [Santaló, 1961], [Santaló, 2004], [Sholander, 1952]. When F is a convex polygon, such as the one in Figure 4.13a, it is easy to see that the equality $K_r = K + Lr + \pi r^2$ holds and when F is a nonconvex polygon, such as the one in Figure 4.13b, that the corresponding inequality holds.

The Kakeya needle problem

In 1917, the Japanese mathematician Soichi Kakeya (1886–1947) asked for the plane region of minimum area in which a line segment of length 1 (the "needle") can be turned through 360°. It can be proved that the convex figure solving the Kakeya needle problem is the equilateral triangle with altitude 1. For many years it was believed that the nonconvex plane region that solves the problem was a deltoid (a hypocycloid with three cusps) with area $\pi/8$. However, a surprising result of A. S. Besicovitch [Besicovitch, 1928] shows that the needle problem has no solution—there are non-convex regions with areas arbitrarily small where the given line segment may still rotate 360°.

For convex regions with area K and perimeter L, we may consider four additional parameters:

(i) the *diameter D*, the largest distance between two points of the region;

(ii) the *width E*, the smallest distance d such that the region can be located between two parallel lines d units apart;

(iii) the *radius R* of the smallest circle containing the region; and

(iv) the *radius r* of the largest circle contained inside the region.

Then we have the following inequalities [Santaló, 1961, 2004]:

$$D^2 \geq 4K/\pi \qquad K \geq E^2/\sqrt{3} \qquad L \geq \pi E \qquad L \leq \pi D$$
$$D \geq E \qquad R \leq D/\sqrt{3} \qquad 4R \leq L \qquad 2r \leq D$$
$$DE \leq 2K \qquad 4K \leq 2EL - \pi E^2 \qquad Lr \leq 2K \qquad K \leq R(L - \pi R)$$

The reader is invited to create pictures illustrating the above inequalities.

Land use laws and inequalities

Local laws concerning land use invariably fix certain requirements for the characteristics that a building must have in relation to its location (e.g., "... 480 m² villa on a lot measuring 1200 m²; land coverage 35%, maximum 2 levels, height 7 m above street level, ..."). In general, such laws fix the limits in terms of inequalities. A typical situation is the case in which the lot is bounded by a curve C, and one needs to determine another curve C' inside of and a fixed distance from C, enclosing an area $K(C')$ which does not exceed $p\%$ of $K(C)$. Moreover, for a building of n floors, inside C' it must be possible to locate a circle of radius $r(C')$ greater than a given value $r(n)$. Thus

$$K(C') \leq \frac{p}{100} K(C) \quad \text{and} \quad r(C') \geq r(n).$$

From the geometric point of view, C' is parallel to C, i.e. from any point on C (except at points with no tangent lines, e.g., corners) one constructs the normal line (perpendicular to the tangent line) and locates a point of C' at a given distance. This transformation is neither an isometry nor a similarity.

4.6 Challenges

4.1 If a, b, c denote the lengths of the sides of a triangle, s the semiperimeter, K its area, R the circumradius and r the inradius, show that:

(a) $abcs \geq 8K^2$,

(b) $abc \geq 8r^2s$,

(c) $2K \leq Rs$,

(d) $2s^2 \geq 27Rr$,

(e) $s^2 \geq 27r^2$.

4.2 In the proof of Lemma 4.4, we showed that $K \leq pq/2$ for a convex quadrilateral Q with diagonals p, q and area K. Establish the stronger result: K is equal to one-half the area of a parallelogram P whose sides are parallel to and equal in length to p and q. (Hint: a visual argument dissects two copies of Q and rearranges the pieces to form P.)

4.3 Let a, b, c denote the lengths of the sides, L the perimeter, and K the area of a triangle. Prove

(a) $K \leq \frac{\sqrt{3}}{4}(abc)^{2/3}$,

(b) the *isoperimetric inequality for triangles*: $L^2 \geq 12\sqrt{3}K$ with equality in each if and only if the triangle is equilateral.

4.4 Prove the *isoperimetric inequality for quadrilaterals*: if L is the perimeter and K the area of a quadrilateral Q, then $L^2 \geq 16K$, with equality if and only if Q is a square. (Note: the isoperimetric inequality for n-gons is $L^2 \geq 4nK/\cot(\pi/n)$.)

4.5 Let Q be a convex quadrilateral with area K and diagonals of length p, q. Prove that $4K \leq p^2 + q^2$, with equality if and only if the diagonals are perpendicular and $p = q$.

4.6 Show that if the words "in the same order" are removed from Theorem 4.7, the conclusion no longer holds. (Hint: given a cyclic n-gon, connect each vertex to the circumcenter and rearrange the resulting segments to construct a *different* cyclic n-gon with the *same* area.)

4.7 *Brahmagupta's formula* for the area K of a cyclic quadrilateral with side lengths a, b, c, and d is $K = \sqrt{(s-a)(s-b)(s-c)(s-d)}$, where s denotes the semi-perimeter $(a + b + c + d)/2$ [Niven, 1981].

(a) Prove that, among all quadrilaterals that can be inscribed in a given circle, the square has the largest area.

(b) Prove that the area K of a bicentric quadrilateral with side lengths a, b, c, and d is $K = \sqrt{abcd}$.

4.8 Let a, b, c be positive real numbers. Show that a, b, c are the sides of a triangle if and only if

$$a^2 + b^2 + c^2 < 2\sqrt{a^2b^2 + b^2c^2 + c^2a^2}.$$

4.9 Create a visual proof that, in the class of right triangles, $R \geq (1+\sqrt{2})r$ with equality for isosceles right triangles.

4.10 Show that for any triangle ABC,

$$h_a h_b + h_b h_c + h_c h_a \leq \left(\frac{a+b+c}{2}\right)^2.$$

(Hint: Use Euler's triangle inequality.)

4.11 Prove the following inequalities for a bicentric quadrilateral with sides a, b, c, d and area K:

(a) $K \leq s^2/4$,

(b) $K \leq (ab + ac + ad + bc + bd + cd)/6$,

(c) $4K^{3/2} \leq abc + abd + acd + bcd$.

4.12 Consider the following five figures formed by segments and arcs of circles, containing in each case six circles of radius 1. Which figure has the minimum perimeter? Which has the maximum perimeter?

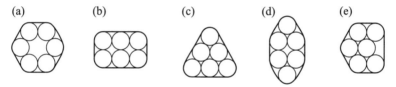

(a)   (b)   (c)   (d)   (e)

Figure 4.14.

An open problem

By the isoperimetric inequality (Theorem 4.2), $L^2 \geq 4\pi K$, so $L^2 \geq 12K$. Find an explicit construction taking twelve pairwise disjoint circles, all of radius r, cutting them into a finite number of parts, and arranging the parts inside a square of side length $2\pi r$ [Alsina, 1992].

CHAPTER 5

Using reflections

An *isometry* in the plane is a transformation that preserves distances. The primary isometries are reflections, rotations, and translations. Preservation of distance implies that isometries preserve angles and areas, so shapes are invariant. As we shall see in this chapter and the next, using isometries is a powerful method for proving geometrical inequalities. Reflections will be used for three different purposes, which we illustrate in the next three examples.

Example 5.1. *Minimal paths*

Given two points A and B on the same side of a line L, what point C in L makes $|AC| + |CB|$ a minimum?

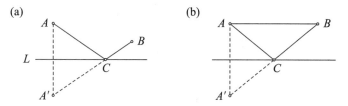

Figure 5.1.

Using L as a mirror, point A reflects to point A', so the line segment $A'B$ is the path of minimum length joining A' and B, and therefore the path of minimum length from A to B containing a point of L is $AC \cup CB$, as illustrated in Figure 5.1a. When the segment AB is parallel to L, as in Figure 5.1b, we have a proof of

Theorem 5.1. *Among all triangles with a specified base and area, the isosceles triangle has minimum perimeter.*

Example 5.2. *Chords dividing a figure into two parts with equal area and equal perimeter*

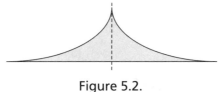

Figure 5.2.

Obviously, whenever we can locate an axis of symmetry in a figure, it determines an equal partition of area and perimeter. This observation, though elementary, is surprisingly useful.

Example 5.3. *Generating new figures with the same perimeter but greater area*

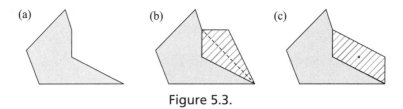

Figure 5.3.

For example, when dealing with concave figures like the region in Figure 5.3a, a reflection of a portion of the boundary in a line (Figure 5.3b) or a point (Figure 5.3c) yields a new region with the same perimeter but greater area. The process can be repeated to produce a convex figure with the same perimeter as the original.

5.1 An inscribed triangle with minimum perimeter

Let P be a fixed point inside an acute angle, and let Q and R be points on the sides of the angle. For what location of Q and R is the perimeter of triangle PQR a minimum? To answer this question we first reflect P in each side of the angle to locate points P' and P'', as shown in Figure 5.4a. Draw $P'P''$, intersecting the sides of the angle at R and Q. We claim triangle PQR has minimum perimeter among all triangles with vertices at P and on the sides of the given angle.

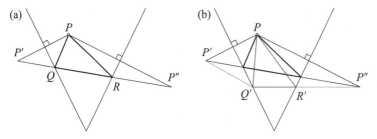

Figure 5.4.

To establish the minimality of the perimeter of triangle PQR, let Q' and R' be any two other points on the sides of the angle, and consider triangle $PQ'R'$, as in Figure 5.4b. Then

$$
\begin{aligned}
|PQ| + |QR| + |RP| = \left|P'Q\right| + |QR| + \left|RP''\right| &= \left|P'P''\right| \\
&\leq \left|P'Q'\right| + \left|Q'R'\right| + \left|R'P''\right| \\
&= \left|PQ'\right| + \left|Q'R'\right| + \left|R'P\right|
\end{aligned}
$$

and consequently triangle PQR has minimum perimeter.

5.2 Altitudes and the orthic triangle

In 1775, Giovanni Francesco Fagnano dei Toschi (1715–1797) posed the following problem: given an acute triangle, find the inscribed triangle of minimum perimeter. By *inscribed triangle* in a given triangle ABC, we mean a triangle PQR such that each vertex P, Q, R lies on a different side of ABC. Fagnano solved the problem using calculus, but we present a non-calculus solution using symmetry due to Lipót Fejér (1880–1959) [Kazarinoff, 1961]. The solution involves the *orthic triangle*—the triangle PQR whose vertices are the feet of the altitudes from each of the vertices of ABC, as illustrated in Figure 5.5a.

Theorem 5.2. *In any acute triangle, the inscribed triangle with minimum perimeter is the orthic triangle.*

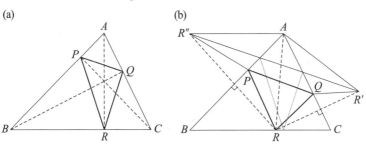

Figure 5.5.

Proof. To find the locations for points P, Q, R that minimize the perimeter of PQR, reflect point R in sides AB and AC (as in the previous section) to locate points R' and R''. Thus the perimeter of PQR is equal to $|R''P| + |PQ| + |QR'|$. The perimeter of PQR will be a minimum whenever R'', P, Q, and R' all lie on the same line. So for any given R, this gives the optimal location for P and Q. To find the optimal location for R, note that triangle $R''AR'$ is isosceles, with $|AR''| = |AR'| = |AR|$ and vertex angle $\angle R''AR' = 2\angle BAC$. Since the size of the vertex angle of triangle $R''AR'$ does not depend on R, the base $R''R'$ (the perimeter of PQR) will be shortest when the legs are shortest, and the legs are shortest when $|AR|$ is a minimum, i.e., when AR is perpendicular to BC.

5.3   Steiner symmetrization

Assume that among all figures with the same fixed perimeter L, there is one figure F with maximum area K. What geometrical characteristics would such a figure F have? Jakob Steiner (1796–1863) proved a variety of results concerning this question (see [Kazarinoff, 1961] for a detailed discussion). We present some of his ideas visually to show that in all cases, axes of symmetry play a crucial role.

Jakob Steiner

Figure 5.6.

Lemma 5.1. *Among all trapezoids with the same parallel bases and altitude, the isosceles trapezoid has minimum perimeter.*

Proof. The trapezoids in Figure 5.7a and 5.7b have the same parallel bases and altitude, and hence the same area; however, the isosceles trapezoid in Figure 5.7b has a vertical axis of symmetry, the perpendicular bisector of

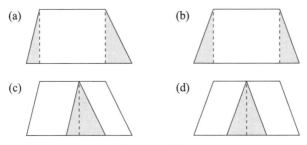

Figure 5.7.

the parallel bases. To conclude that the trapezoid in Figure 5.7b has the smaller perimeter, we need only show that the sum of the lengths of the non-parallel sides is smaller. Rearranging the parts of the trapezoids and applying Theorem 5.1 to the gray triangles in Figures 5.7 and 5.7d accomplishes that.

One of Steiner's ingenious proofs of the isoperimetric inequality uses the following theorem [Kazarinoff, 1961].

Theorem 5.3. *Among all convex figures with a given perimeter, one with maximum area has an axis of symmetry in every direction.*

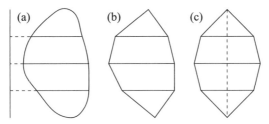

Figure 5.8.

Proof. Slice the convex figure into thin parallel strips with bases perpendicular to the given direction (Figure 5.8a). Now we can construct a new convex figure made up of trapezoids (Figure 5.8b). Applying Lemma 5.1 we can construct another figure made up of isosceles trapezoids with the same bases and area, but a smaller perimeter, and lined up so they have a common perpendicular bisector (Figure 5.8c). If we slice the original figure into thinner and thinner strips, the approximating polygons have areas and perimeters approaching those of the original figure.

5.4   Another minimal path

In Example 5.1 we saw how to find the minimal path from one point to another touching a given line not passing between the points. As we noted, the given line acts like a mirror and the path like a ray of light. This lighting principle is a key tool for finding minimal paths subject to given restraints.

A classical geometric problem is the following: given two points A and B in the interior of an angle determined by two lines r and s, what is the minimal path from point A to r to s to point B?

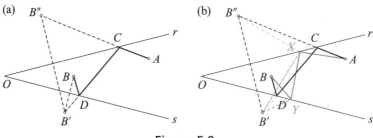

Figure 5.9.

The key is multiple reflections. As illustrated in Figure 5.9a, let B' be the reflection of B in line s and B'' the reflection of B' in line r. Let AB'' intersect r at C and let CB' intersect s at D. Then C and D are the required points on r and s respectively so $ACDB$ is the path from A to r to s to B of minimum length, and that length is

$$|AC|+|CD|+|DB| = |AC|+|CD|+\left|DB'\right| = |AC|+\left|CB'\right| = \left|AB''\right|.$$

To verify the minimality of the length of this path, let X and Y be any other two points on r and s, respectively, as illustrated in Figure 5.9b. Then

$$\left|AB''\right| \le |AX| + \left|XB''\right| = |AX| + \left|XB'\right| \le |AX| + |XY| + \left|YB'\right|$$
$$= |AX| + |XY| + |YB|.$$

One might also ask for the shortest path from A to s to r to B, as seen in Figure 5.10. This is constructed analogously, and may be greater than, equal to, or shorter than the path in Figure 5.9.

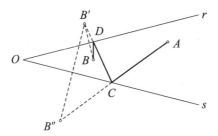

Figure 5.10.

It can be shown [Courant and Robbins, 1941] that the first path is shorter than the second when O and C lie on the same side of the line through A and B, extended to intersect the angle.

5.5 An inscribed triangle with minimum area

Another classical geometric problem is the following: given an angle that is less than 180° and a point M in its interior, construct the line segment through M that intersects the sides of the angle to form a triangle with minimum area.

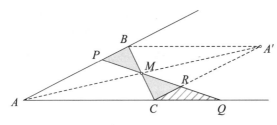

Figure 5.11.

Let A' be the point symmetric to A with respect to M, and construct segments from A' to the sides of the angle to form a parallelogram $ABA'C$ whose diagonals BC and AA' meet at M (see Figure 5.11). Then the triangle with minimum area is ABC. Any other triangle APQ with PQ passing through M would have additional area RCQ since the triangles MBP and MCR are symmetric with respect to M and hence congruent.

5.6 Challenges

5.1 Given a triangle ABC, reflect it in the line perpendicular to the bisector of angle C to form a trapezoid $AA'B'B$, as illustrated in

Figure 5.12. Use this figure to show that $w_c \leq 2ab/(a + b)$, establishing an interesting inequality between the length of the angle bisector and the harmonic mean of the two sides adjoining the angle.

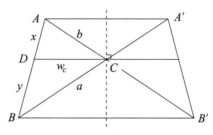

Figure 5.12.

5.2 Two points A and B lie on the same side of a straight line ℓ. Find the minimal path joining A to B containing some segment of ℓ of length s.

5.3 Let P be a point inside an acute angle, and let Q and R be variable points, one on each side of the angle. Find the positions of Q and R so triangle PQR has minimum perimeter.

5.4 If $a, b, c > 0$ determine a triangle and ℓ is a given line containing side a, how many triangles with base a can one draw?

5.5 Does Theorem 5.2 (Fagnano's problem) have an analog for right or obtuse triangles?

CHAPTER **6**

Using rotations

Like reflections, rotations are useful isometries, since in addition to preserving distances and angles they preserve orientation. As we shall see in this chapter, rotating figures is a useful way to create a visual explanation of an inequality.

In rotation, a figure or portion of a figure is rotated in the plane about a given point through a given angle, as illustrated in the following example.

Example 6.1. *A maximal distance problem*

Let ABC be an equilateral triangle whose sides have length a. Suppose a point P is located a fixed distance u from vertex A, and a fixed distance v from vertex B (where $|u - v| < a$), as illustrated in Figure 6.1a. What is the upper bound on the distance between P and vertex C?

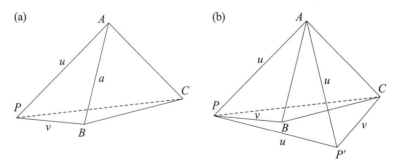

Figure 6.1.

Rotate triangle APB $60°$ counterclockwise about vertex A to create triangle $AP'C$, as illustrated in Figure 6.1b. Since $|PA| = |P'A| = u, |PP'| = u$, and thus $|PC| \leq |PP'| + |P'C| = u + v$.

In Challenge 6.1, you will show that we have equality (i.e., $|PC| = |PA| + |PB|$) when P lies on a portion of the circumcircle of ABC.

6.1 Ptolemy's inequality

A classical result of Claudius Ptolemy of Alexandria (circa 85–165), known
as *Ptolemy's theorem*, states that for a cyclic quadrilateral with side lengths
a, b, c, d (in that order) and diagonals of lengths p and q, the product of the
lengths of the diagonals equals the sum of the products of the lengths of the
opposite sides, $pq = ac + bd$. For a general convex quadrilateral, we have
Ptolemy's inequality:

Theorem 6.1. *For a convex quadrilateral with sides of length a, b, c, d (in
that order) and diagonals of length p and q, we have $pq \leq ac + bd$.*

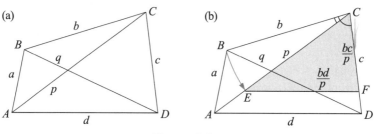

Figure 6.2.

Proof. We start with the convex quadrilateral $ABCD$ in Figure 6.2a. In
Figure 6.2b, we rotate side BC about vertex B to locate point E so $|BC| =
|EC| = b$, and draw EF parallel to AD. This creates triangle CEF (in
gray) similar to triangle CAD, with side lengths as indicated.

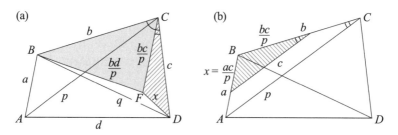

Figure 6.3.

In Figure 6.3a, we rotate the gray triangle about vertex C as shown, and
let $x = |FD|$, and thus $bd/p + x \geq q$. We now rotate and translate triangle
DFC (the striped triangle) to the position shown in Figure 6.3b, and since

it is similar to triangle ABC (because of the proportional sides and equal included angles), $x = ac/p$. Thus $bd/p + ac/p \geq q$, and the inequality follows.

When $ABCD$ is cyclic, $\angle CBD = \angle CAD = \angle CEF$ in Figures 6.2a and 6.2b, so $\angle CBD = \angle CBF$ in Figure 6.3a, and consequently the line segments BF and FD lie on BD. We then have equality rather than inequality, so Ptolemy's theorem follows from the proof of Ptolemy's inequality.

6.2 Fermat's problem for Torricelli

Given a triangle ABC (see Figure 6.5a), where is the point F such that the sum $|FA| + |FB| + |FC|$ of the distances from F to the three vertices is a minimum? This old problem is often referred to as *Fermat's problem for Torricelli*, and the point F is often called the *Fermat point* of the triangle. Pierre Fermat (1601–1665) gave the problem to Evangelista Torricelli (1608–1647), and Torricelli solved it in several ways. It is also known as *Steiner's problem*. If ABC contains an angle of 120° or more, the Fermat point is the vertex of this obtuse angle, so we will only consider triangles in which each angle measures less than 120°.

Evangelista Torricelli   Pierre Fermat

Figure 6.4.

There is a simple way to locate the Fermat point of such a triangle. Construct equilateral triangles on the sides of triangle ABC, and join each vertex of triangle ABC to the exterior vertex of the opposite equilateral triangle. Those three lines intersect at the Fermat point F, as illustrated in Figure 6.5b.

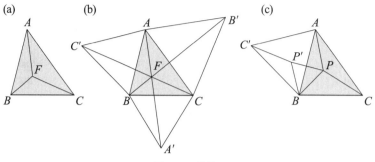

Figure 6.5.

Why does this work? The following proof [Niven, 1981] uses rotational isometry. In Figure 6.5c, we select any point P inside triangle ABC and connect P to the three vertices. Now rotate triangle APB 60° counterclockwise about vertex B to form triangle $C'P'B$ and draw the lines $C'A$ and $P'P$. Now triangle ABC' is equilateral, $|PA| = |C'P'|, |PB| = |P'B|$, and consequently triangle BPP' is equilateral, so $|PB| = |P'P|$. Thus

$$|PA| + |PB| + |PC| = |C'P'| + |P'P| + |PC|.$$

This last sum will be a minimum when P and P' both lie on the line $C'C$ (because C' is the image of A under the rotation, its position does not depend on P). Thus $|PA| + |PB| + |PC|$ is a minimum if and only if P lies on $C'C$, and for such a P, $\angle BPC' = 60°$. Since the choice of which side of the triangle to rotate was arbitrary, P must also lie on $B'B$ and $A'A$.

6.3   The Weitzenböck and Hadwiger-Finsler inequalities

Problem 2 on the Third International Mathematical Olympiad in 1961 was [Djukić et al, 2006]

Let a, b, c be the sides of a triangle, and K its area. Prove:

$$a^2 + b^2 + c^2 \geq 4\sqrt{3}K.$$

In what case does equality hold?

This inequality is known in the literature (e.g., see [Steele, 2004]) as *Weitzenböck's inequality* (sometimes spelled Weizenbock) from a paper published in 1919 by R. Weitzenböck in *Mathematische Zeitschrift*. Many analytical proofs of the inequality are known.

There is a geometrical interpretation of this inequality that seems to have been overlooked. If we multiply both sides of the inequality by $\sqrt{3}/4$, then it can be written as

$$K_a + K_b + K_c \geq 3K,$$

where K_x denotes the area of an equilateral triangle with side length x. That is, the sum of the areas K_a, K_b, and K_c of the three equilateral triangles shaded gray in Figure 6.6 is at least three times the area K of the unshaded triangle.

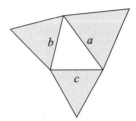

Figure 6.6.

We now present a purely geometric proof [Alsina and Nelsen, 2008] of Weitzenböck's inequality using the Fermat point of the original triangle. As we saw in Section 6.2, the Fermat point of a triangle ABC is the point F in or on the triangle for which the sum $|FA| + |FB| + |FC|$ is a minimum. When each of the angles of the triangle is smaller than $120°$, the point F is the point of intersection of the lines connecting the vertices A, B, and C to the vertices of equilateral triangles constructed outwardly on the sides of the triangle. When one of the vertices of triangle ABC measures $120°$ or more, then that vertex is the Fermat point.

We first consider the case where each angle of the triangle is less than $120°$. Let x, y, and z denote the lengths of the line segments joining the Fermat point F to the vertices, as illustrated in Figure 6.7a, and note that the two acute angles in each triangle with a vertex at F sum to $60°$. Hence three copies of the triangle shaded gray plus an unshaded equilateral triangle whose sides measure $|x - y|$ form an equilateral triangle of area K_c. The same is true of the other triangles sharing the vertex F, and hence

$$K_a + K_b + K_c = 3K + K_{|x-y|} + K_{|y-z|} + K_{|z-x|} \geq 3K, \qquad (6.1)$$

which establishes the inequality in this case since $K_{|x-y|}, K_{|y-z|}$, and $K_{|z-x|}$ are each nonnegative.

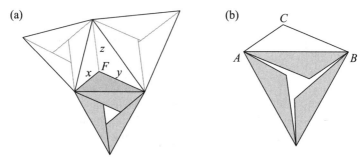

Figure 6.7.

When one angle (say C) measures $120°$ or more, then, as illustrated in Figure 6.7b, we have

$$K_a + K_b + K_c \geq K_c \geq 3K, \tag{6.2}$$

which completes the proof.

It follows from our proof that we have equality in Weitzenböck's inequality if and only if $x = y = z$, so the three triangles with a common vertex at F are congruent and hence $a = b = c$, i.e., the original triangle is equilateral.

The relationship in (6.1) is actually stronger than the Weitzenböck inequality, and enables us to now prove another inequality, the *Hadwider-Finsler inequality* [Bottema et al., 1968; Steele, 2004]: If a, b, and c are the lengths of the sides of a triangle with area K, then

$$a^2 + b^2 + c^2 \geq 4\sqrt{3}K + (a - b)^2 + (b - c)^2 + (c - a)^2.$$

In terms of areas of triangles, this is equivalent to

$$K_a + K_b + K_c \geq 3K + K_{|a-b|} + K_{|b-c|} + K_{|c-a|}. \tag{6.3}$$

To prove that (6.1) implies (6.3) when all three angles measure less than $120°$, we need only show that $|x - y| \geq |a - b|$, $|y - z| \geq |b - c|$, and $|z - x| \geq |c - a|$. Without loss of generality, assume that $a \geq b \geq c$. Within triangle ABC, reflect triangle ACF in the segment CF as shown in Figure 6.8 to create two congruent light gray triangles (recall that each of the three angles at F measures $120°$). Then in the dark gray triangle we have $b + y - x \geq a$, or $y - x \geq a - b$. The other two inequalities are established similarly, and hence from (6.1) we have

$$K_a + K_b + K_c = 3K + K_{|x-y|} + K_{|y-z|} + K_{|z-x|}$$

$$\geq 3K + K_{|a-b|} + K_{|b-c|} + K_{|c-a|},$$

which completes the proof in this case.

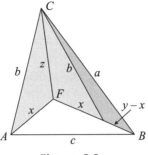

Figure 6.8.

When one angle (say C) measures $120°$ or more, we have $z = 0$, $x = b$, and $y = a$. We refine the inequality in (6.2) to $K_a + K_b + K_c \geq 3K + K_{|a-b|} + K_a + K_b$ (see Figure 6.7b), and note that $a \geq |b - c|$, and $b \geq |c - a|$, from which the Hadwiger-Finsler inequality follows.

6.4   A maximal chord problem

Suppose two circles intersect in a pair of points A and D. Of all possible line segments passing through A with endpoints on the two circles, which is the longest?

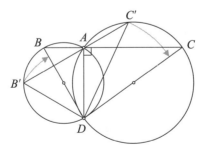

Figure 6.9.

We claim the longest such segment is the line segment BC perpendicular to AD at A, as illustrated in Figure 6.9. Let $B'C'$ be another segment passing through A, and draw $B'D$ and $C'D$. Since $\angle C'B'D = \angle CBD$ and $\angle B'C'D = \angle BCD$, triangle $B'C'D$ is similar to triangle BCD. However, a rotation of triangle $B'C'D$ about vertex D shows that $|C'D| \leq |CD|$ (CD is a diameter), and hence $|B'C'| \leq |BC|$.

6.5 The Pythagorean inequality and the law of cosines

The Pythagorean theorem relates the squares of the lengths of the sides of a right triangle in an equation. The *Pythagorean inequality* relates the squares of the lengths of the sides of an acute triangle in an inequality: if a, b, c are the lengths of the sides of an acute triangle, then $c^2 \leq a^2 + b^2$ (and $b^2 \leq c^2 + a^2$ and $a^2 \leq b^2 + c^2$, but we will prove only the first inequality).

In triangle ABC, draw the altitude to side BC as shown in Figure 6.10a, and rotate that line segment clockwise about vertex A. The length of the altitude is $b \sin C$, so we draw squares with side lengths $b \sin C$ and $a - b \cos C$.

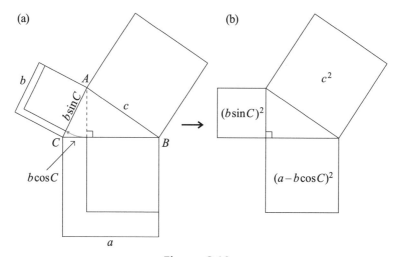

Figure 6.10.

Those two squares and the one with area c^2 are the squares of the lengths of the sides of a right triangle, as seen in Figure 6.10b, so by the Pythagorean theorem,

$$c^2 = (b \sin C)^2 + (a - b \cos C)^2. \tag{6.4}$$

But $b \sin C \leq b$ and $a - b \cos C \leq a$, from which the inequality follows. We have equality if and only if $\sin C = 1$ and $\cos C = 0$, i.e., when C is a right angle.

As a bonus, we obtain the law of cosines by simplifying the right-hand side of (6.4) [Sipka, 1988]: $c^2 = a^2 + b^2 - 2ab \cos C$ (and similarly $a^2 = b^2 + c^2 - 2bc \cos A$ and $b^2 = a^2 + c^2 - 2ac \cos B$).

Application 6.1. *Inequalities derived from equalities*

Finding bounds for a term in an equality is a natural method for deriving inequalities. To illustrate this, consider the following problem: If a, b, and c are the side lengths of a triangle, can a^2, b^2, and c^2 be the side lengths of another triangle? In general the answer is no. If ABC is a right triangle with hypotenuse c, then $c^2 = a^2 + b^2$, so a^2, b^2, and c^2 do not determine a triangle. However, for acute triangles, the answer is yes, as a consequence of the Pythagorean inequality. As a second example, we use the law of cosines to prove

Theorem 6.2. *In a triangle ABC,* $\cos A + \cos B + \cos C > 1$.

Proof. From the law of cosines, we have $\cos A = (b^2 + c^2 - a^2)/2bc$, $\cos B = (c^2 + a^2 - b^2)/2ac$, and $\cos C = (a^2 + b^2 - c^2)/2ab$, and hence

$$\cos A + \cos B + \cos C - 1$$
$$= \frac{b^2 + c^2 - a^2}{2bc} + \frac{c^2 + a^2 - b^2}{2ac} + \frac{a^2 + b^2 - c^2}{2ab} - 1$$
$$= \frac{1}{2abc}\left[a(b^2 + c^2 - a^2) + b(c^2 + a^2 - b^2) + c(a^2 + b^2 - c^2) - 2abc\right]$$
$$= \frac{1}{2abc}\left[(-a + b + c)(a - b + c)(a + b - c)\right]$$
$$= \frac{4K^2}{sabc} = \frac{r}{R} > 0,$$

where we have employed Heron's formula $K = \sqrt{s(s-a)(s-b)(s-c)}$ (see Application 4.1) and $K = rs = abc/4R$ (see Lemmas 4.1 and 4.3). In Applications 7.1 and 8.9 we will prove the companion result $\cos A + \cos B + \cos C \le 3/2$.

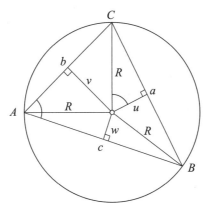

Figure 6.11.

There is a geometric interpretation of the inequality in Theorem 6.2. In triangle ABC with circumradius R, let u, v, and w denote the distances from the circumcenter to the sides, as illustrated in Figure 6.11. Since the angle marked $\measuredangle$ at the circumcenter has the same measure as $\angle A$, we have $\cos A = u/R$, and similarly $\cos B = v/R$ and $\cos C = w/R$. Hence we have proved

Corollary 6.1. *If u, v, w denote the distances from the circumcenter to the sides of a triangle with circumradius R, then $u + v + w > R$.*

6.6 Challenges

6.1 Show that $|PC| = |PA| + |PB|$ in Example 6.1 when P lies on a portion of the circumcircle of ABC. (Hint: Show that equality holds if and only if $APBC$ is a cyclic quadrilateral.)

6.2 Let K denote the area of a quadrilateral with side lengths a, b, c, and d. In Challenge 2.4d, we proved $4K \leq a^2 + b^2 + c^2 + d^2$ algebraically. Using Figure 6.12, create a visual proof of this inequality. (Hint: See Figure 4.8.)

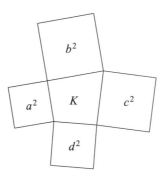

Figure 6.12.

6.3 Consider a convex pentagon with area K that has the property that there is an interior point P in the pentagon such that the five line segments joining P to the vertices meet at $72°$ angles at P. Let $K_i, i = 1, 2, 3, 4, 5$ denote the areas of the five regular pentagons constructed outwardly on the five sides of the given pentagon. Show that $K_1 + K_2 + K_3 + K_4 + K_5 \geq 5K$.

6.4 Given a point P between two parallel lines r and s, find the square of maximum area that lies between the lines and that has a vertex at P.

6.5 Derive Weitzenböck's inequality by establishing the following chain of inequalities:

$$a^2 + b^2 + c^2 \geq ab + bc + ca \geq a\sqrt{bc} + b\sqrt{ca} + c\sqrt{ab}$$
$$\geq 3(abc)^{2/3} \geq 4K\sqrt{3}.$$

6.6 (a) Show that Weitzenböck's inequality is equivalent to

$$a^2b^2 + b^2c^2 + c^2a^2 \leq a^4 + b^4 + c^4.$$

(Hint: Use Heron's formula.)

(b) Prove that $a^2b^2 + b^2c^2 + c^2a^2 \leq a^4 + b^4 + c^4$ actually holds for any real numbers a, b, and c.

6.6 Consider the following extension of Application 6.1. Let a, b, and c be the side lengths of a triangle with the property that for any positive integer n, a^n, b^n, and c^n can also be the side lengths of a triangle. Prove that the original triangle is isosceles [Gelca and Andreescu, 2007].

CHAPTER 7

Employing non-isometric transformations

In the two previous chapters, we examined how reflections and rotations (isometric transformations) may be used in a visual approach to inequalities. Non-isometric transformations—transformations that do not necessarily preserve lengths—constitute an interesting class of mappings for proving some inequalities. We consider three types of non-isometric transformations in this chapter: similarity of figures, measure-preserving transformations, and projections. The first preserves shapes, but changes measures by a given factor or its powers while the others may change shapes of figures but preserve other properties.

7.1 The Erdős-Mordell theorem

In 1935, the following problem proposal appeared in the Advanced Problems section of the *American Mathematical Monthly* [Erdős, 1935]:

3740. *Proposed by Paul Erdős, The University, Manchester, England.*
From a point O inside a given triangle ABC the perpendiculars OP, OQ, OR are drawn to its sides. Prove that

$$|OA| + |OB| + |OC| \geq 2\left(|OP| + |OQ| + |OR|\right).$$

Trigonometric solutions by Mordell and Barrow appeared in [Mordell and Barrow, 1937]; the proofs, however, were not elementary. In fact, no "simple and elementary" proof of what had become known as the *Erdős-Mordell theorem* was known as late as 1956 [Steensholt, 1956]. Since then a variety of proofs have appeared, each in some sense simpler or more elementary than the preceding ones. In 1957 Kazarinoff published a proof [Kazarinoff, 1957] based upon a theorem in Pappus of Alexandria's *Mathematical Collection*;

and a year later Bankoff published a proof [Bankoff, 1958] using orthogonal projections and similar triangles. Proofs using area inequalities appeared in [Komornik, 1997] and [Dergiades, 2004]. Proofs employing Ptolemy's theorem appeared in [Avez, 1993] and [Lee, 2001]. The following visual proof appeared in [Alsina and Nelsen, 2007b].

In Figure 7.1a we see the triangle as described by Erdős, and in Figure 7.1b we denote the lengths of relevant line segments by lower case letters. In terms of that notation, the Erdős-Mordell inequality becomes $x + y + z \geq 2(u + v + w)$.

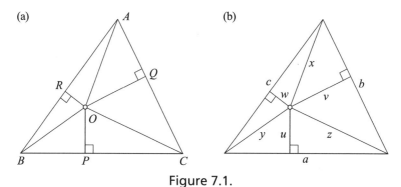

Figure 7.1.

Using similar figures, we construct a trapezoid in Figure 7.2b from three triangles—one similar to *ABC*, the other two similar to the shaded triangles in Figure 7.2a, to prove Lemma 7.1. Our proof applies to acute triangles; when the triangle is obtuse, an analogous construction using similar triangles yields the same inequalities.

Lemma 7.1. *For the triangle ABC in Figure 7.1b, we have $ax \geq bw + cv$, $by \geq aw + cu$, and $cz \geq av + bu$.*

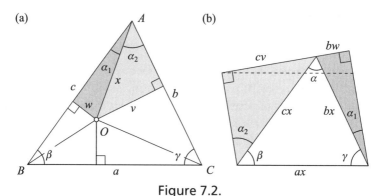

Figure 7.2.

Proof. The dashed line segment in Figure 7.2b has length ax, and hence the Pythagorean comparison yields $ax \geq bw + cv$. The other two inequalities are established analogously.

We should note that the object in Figure 7.2b really is a trapezoid: the three angles at the point where the three triangles meet measure $(\pi/2) - \alpha_2, \alpha = \alpha_1 + \alpha_2$, and $(\pi/2) - \alpha_1$, and thus sum to π.

We now prove

The Erdős-Mordell Theorem 7.1. *If O is a point within a triangle ABC whose distances to the sides are u, v, and w and whose distances to the vertices are x, y, and z, then*

$$x + y + z \geq 2(u + v + w).$$

Proof. From Lemma 7.1 we have $x \geq \frac{b}{a}w + \frac{c}{a}v$, $y \geq \frac{a}{b}w + \frac{c}{b}u$, and $z \geq \frac{a}{c}v + \frac{b}{c}u$; adding these three inequalities yields

$$x + y + z \geq \left(\frac{b}{c} + \frac{c}{b}\right)u + \left(\frac{a}{c} + \frac{c}{a}\right)v + \left(\frac{a}{b} + \frac{b}{a}\right)w. \qquad (7.1)$$

But the AM-GM inequality insures that the coefficients of u, v, and w are each at least 2, from which the desired result follows.

Note 1. The three inequalities in Lemma 7.1 are equalities if and only if O is the center of the circumscribed circle of ABC. This follows from the observation that the trapezoid in Figure 7.3b is a rectangle if and only if $\beta + \alpha_2 = \pi/2$ and $\gamma + \alpha_1 = \pi/2$ (and similarly in the other two cases), so $\angle AOQ = b = \angle COQ$. Hence the right triangles AOQ and COQ are congruent, so $x = z$. Similarly one can show that $x = y$, hence $x = y = z$ so

O must be the circumcenter of ABC. The coefficients of u, v, and w in (7.1) are equal to 2 if and only if $a = b = c$; and consequently we have equality in the Erdős-Mordell inequality if and only if ABC is equilateral and O is its center.

Note 2. Many other inequalities relating x, y, and z to u, v, and w can be derived. For example, applying the AM-GM inequality to the right sides of the inequalities in Lemma 7.1 yields

$$ax \geq 2\sqrt{bcvw}, \quad by \geq 2\sqrt{acuw}, \text{ and } cz \geq 2\sqrt{abuv}.$$

Multiplying these three inequalities together and simplifying yields

$$xyz \geq 8uvw. \qquad (7.2)$$

In triangle ABC in Figure 7.1, let K denote the area, R the circumradius, and h the altitude to side AB. In the proof of Lemma 4.1 we showed that $ab = h(2R)$, i.e., the product of the lengths of any two sides is equal to the length of the altitude to the third side times the diameter of the circumcircle. As a consequence we have

Theorem 7.2. *For the triangle ABC in Figure 7.1, we have*

$$\frac{xy}{w} + \frac{yz}{u} + \frac{xz}{v} \geq 2(x + y + z) \geq 4(u + v + w). \tag{7.3}$$

Proof. Construct triangle $A'B'C'$ by drawing $A'B'$ perpendicular to OC at C, $B'C'$ perpendicular to OA at A, and $A'C'$ perpendicular to OB at B as shown in Figure 7.3. Quadrilateral $OCA'B$ is cyclic since its angles at B and C are right angles, and its circumcircle (which is also the circumcircle of triangle BOC) has diameter OA'. Hence, from the statement preceding this theorem, $yz = u \cdot |OA'|$, or $|OA'| = yz/u$. Similarly, $|OB'| = xz/v$ and $|OC'| = xy/w$. Applying the Erdős-Mordell inequality to triangle $A'B'C'$ yields

$$|OA'| + |OB| + |OC'| \geq 2(x + y + z),$$

i.e., (7.3) holds.

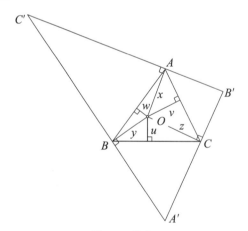

Figure 7.3.

Inequalities similar to (7.1)–(7.3) can be similarly derived. Examples [Oppenheim, 1961; Ehret, 1970] are the following:

Corollary 7.1. *For the triangle ABC in Figure 7.1, we have*

$$ux + vy + wz \geq 2(uv + vw + wu), \tag{7.4}$$

$$\frac{1}{u} + \frac{1}{v} + \frac{1}{w} \geq 2\left(\frac{1}{x} + \frac{1}{y} + \frac{1}{z}\right). \tag{7.5}$$

$$xy + yz + zx \geq 2(ux + vy + wz), \tag{7.6}$$

$$xy + yz + zx \geq 4(uv + vw + wu). \tag{7.7}$$

Proof. From Figure 7.1b, the altitude to side BC is less than or equal to $x + u$, hence the area K of triangle ABC satisfies $K \leq a(x+u)/2$. But $K = (au + bv + cw)/2$, and consequently $ax \geq bv + cw$, or $ux \geq (b/a)uv + (c/a)uw$. Similarly, $vy \geq (a/b)uv + (c/b)vw$ and $wz \geq (a/c)uw + (b/c)vw$. Thus

$$ux + vy + wz \geq \left(\frac{a}{b} + \frac{b}{a}\right)uv + \left(\frac{b}{c} + \frac{c}{b}\right)vw + \left(\frac{a}{c} + \frac{c}{a}\right)uw$$
$$\geq 2(uv + vw + wu),$$

which establishes (7.4). Applying (7.4) to triangle $A'B'C'$ in Figure 7.3 yields

$$x\left|OA'\right| + y\left|OB'\right| + z\left|OC'\right| \geq 2(xy + yz + xz).$$

Substituting $|OA'| = yz/u$, $|OB'| = xz/v$ and $|OC'| = xy/w$ now gives

$$xyz\left(\frac{1}{u} + \frac{1}{v} + \frac{1}{w}\right) \geq 2xyz\left(\frac{1}{x} + \frac{1}{y} + \frac{1}{z}\right),$$

which establishes (7.5). Now apply (7.5) to triangle $A'B'C'$ in Figure 7.3 to get

$$\frac{1}{x} + \frac{1}{y} + \frac{1}{z} \geq 2\left(\frac{1}{|OA'|} + \frac{1}{|OB'|} + \frac{1}{|OC'|}\right) = 2\left(\frac{u}{yz} + \frac{v}{xz} + \frac{w}{xy}\right)$$

and multiply by xyz to obtain (7.6). Finally, (7.7) follows from (7.4) and (7.6). ∎

Application 7.1. *Trigonometric inequalities derived from the Erdős-Mordell theorem and its consequences* [Ehret, 1970]

We now let O be the incenter of triangle ABC, so $u = v = w = r$ is the inradius of ABC. We now find inequalities for the sum and product of the sines and of the cosines of A, B, C and $A/2$, $B/2$, $C/2$.

Since $r = x\sin(A/2) = y\sin(B/2) = z\sin(C/2)$, employing (7.5) yields

$$\frac{3}{r} \geq \frac{2}{r}\left(\sin(A/2) + \sin(B/2) + \sin(C/2)\right),$$

or

$$\sin(A/2) + \sin(B/2) + \sin(C/2) \le 3/2.$$

Similarly, (7.2) yields

$$\sin(A/2)\sin(B/2)\sin(C/2) \le 1/8.$$

But for any triangle, $\cos A + \cos B + \cos C = 1 + 4\sin(A/2)\sin(B/2) \times \sin(C/2)$, and hence

$$\cos A + \cos B + \cos C \le 3/2. \tag{7.8}$$

Applying the AM-GM inequality to (7.8) produces

$$\cos A \cos B \cos C \le 1/8.$$

Substituting $\cos A = 2\cos^2(A/2) - 1$ (and similarly for $\cos B$ and $\cos C$) and applying the arithmetic mean-root mean square inequality in (7.8) yields

$$\cos(A/2) + \cos(B/2) + \cos(C/2)$$
$$\le 3\sqrt{\frac{\cos^2(A/2) + \cos^2(B/2) + \cos^2(C/2)}{3}}$$
$$= 3\sqrt{\frac{3 + \cos A + \cos B + \cos C}{6}} \le \frac{3\sqrt{3}}{2},$$

and applying the AM-GM inequality to the above yields

$$\cos(A/2)\cos(B/2)\cos(C/2) \le \frac{3\sqrt{3}}{8}.$$

Finally, since $\sin A + \sin B + \sin C = 4\cos(A/2)\cos(B/2)\cos(C/2)$, we have

$$\sin A + \sin B + \sin C \le \frac{3\sqrt{3}}{2},$$

and using the AM-GM inequality once more yields

$$\sin A \sin B \sin C \le \frac{3\sqrt{3}}{8}.$$

We have equality in each inequality above if and only if the triangle is equilateral. Many similar inequalities can be established, see Challenge 7.1.

Application 7.2. *Carnot's theorem*

There are instances where the proof of an inequality yields as a special case the proof of an equality. Such is the case with Lemma 7.1 when the point O is the circumcenter. In that case we obtain a proof of the following theorem, due to Lazare Nicolas Marguérite Carnot (1753–1823).

Carnot's Theorem 7.3. *In any acute triangle, the sum of the distances from the circumcenter to the sides is equal to the sum of the inradius and circumradius.*

Proof. As we observed in Note 1 following the proof of the Erdős-Mordell theorem, when O is the circumcenter the inequalities in Lemma 7.1 are equalities, with $x = y = z = R$. Then we have $aR = bw + cv$, $bR = aw + cu$, and $cR = av + bu$. From Figure 7.1 $(au + bv + cw)/2 = K$, and from Lemma 4.3 $K = rs = r(a + b + c)/2$, so

$$(a+b+c)(u+v+w) = (au+bv+cw)+(aw+cu)+(av+bu)+(bw+cv)$$
$$= 2K + (a + b + c)R$$
$$= (a + b + c)(r + R).$$

Hence $u + v + w = r + R$. The theorem can be extended to include obtuse triangles by using signed distances from O to the sides; see [Honsberger, 1985].

L. N. M. Carnot   Paul Erdős

Figure 7.4.

7.2 Another Erdős triangle inequality

Another inequality for line segments in a triangle comes from the following problem, attributed to Paul Erdős in [Kazarinoff, 1961]:

Let P be any point inside a triangle ABC, and suppose that AP, BP, and CP extended meet the sides at A', B', and C', respectively. Prove that $|PA'| + |PB'| + |PC'|$ is less than the length of the longest side of the triangle.

Before proving the inequality in the problem, we prove a lemma:

Lemma 7.2. *In any triangle, the distance from a vertex to a point on the opposite side is less than the longest side of the triangle.*

Proof. Using the notation in Figure 7.5a, we want to show that $t < \max(a, b, c)$.

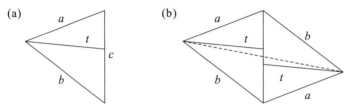

Figure 7.5.

Duplicating the triangle to form the parallelogram in Figure 7.5b shows that $2t < a + b$, and hence

$$t < \frac{a+b}{2} \leq \max(a, b) \leq \max(a, b, c).$$

To prove the assertion in the problem, we need to show that $x + y + z < \max(a, b, c)$ in Figure 7.6a. Draw lines through P parallel to the sides of the triangle, partitioning it into three parallelograms and three triangles (shaded gray in Figure 7.6b) similar to triangle ABC. If we denote the lengths of the subsegments of AB by u, v, and w, then the lengths of the sides of the gray triangle containing the segment of length x are w, bw/c, and aw/c, the lengths of the sides of the gray triangle containing the segment of length y are u, bu/c, and au/c, and the lengths of the sides of the gray triangle containing the segment of length z are v, bv/c, and av/c.

Hence, using Lemma 7.2, we have

$$x \leq \max(w, aw/c, bw/c), \quad y \leq \max(u, au/c, bu/c),$$
$$z \leq \max(v, av/c, bv/c),$$

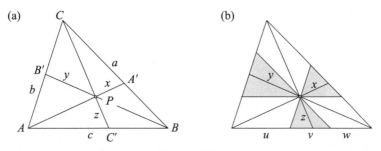

Figure 7.6.

and adding the inequalities yields

$$x + y + z \le \max\left(w, \frac{aw}{c}, \frac{bw}{c}\right) + \max\left(u, \frac{au}{c}, \frac{bu}{c}\right) + \max\left(v, \frac{av}{c}, \frac{bv}{c}\right)$$

$$= \frac{w}{c}\max(c, a, b) + \frac{u}{c}\max(c, a, b) + \frac{v}{c}\max(c, a, b)$$

$$= \left(\frac{w}{c} + \frac{u}{c} + \frac{v}{c}\right)\max(c, a, b) = \max(c, a, b).$$

7.3   The Cauchy-Schwarz inequality

In 1821 Augustin-Louis Cauchy (1789–1857) published the following in-equality that now bears his name: for any two sets $a_1, a_2, \ldots, a_n$ and $b_1, b_2, \ldots, b_n$ of real numbers, we have

$$|a_1b_1 + a_2b_2 + \cdots + a_nb_n| \le \sqrt{a_1^2 + a_2^2 + \cdots + a_n^2}\sqrt{b_1^2 + b_2^2 + \cdots + b_n^2},$$

$$(7.9)$$

with equality if and only if the two sets are proportional, i.e., $a_ib_j = a_jb_i$ for all i and j between 1 and n. Integral versions of this inequality were published by Victor Yacovlevich Bunyakovski (1804–1889) in 1859 and Hermann Amandus Schwarz (1843–1921) in 1885. Today the inequality is known either as the Cauchy-Schwarz inequality or the Cauchy-Bunyakovski-Schwarz inequality.

For $n = 2$, (7.9) is equivalent to

$$(a^2 + b^2)(x^2 + y^2) \ge (ax + by)^2. \qquad (7.10)$$

One of the simplest algebraic proofs of (7.10) follows from Diophantus (circa 200–284) of Alexandria's sum of squares identity:

$$(a^2 + b^2)(x^2 + y^2) = (ax + by)^2 + (ay - bx)^2.$$

Augustin-Louis Cauchy   Hermann Amandus Schwarz

Figure 7.7.

This equality is easily established by algebra, and since $(ay - bx)^2 \geq 0$, (7.10) follows immediately. Diophantus used his identity to show that the product of any two integers that can be written as a sum of squares of integers is again a sum of integer squares. For a visual proof of Diophantus' identity using non-isometric transformations, see [Nelsen, 1993].

We now present three geometric proofs and another algebraic proof of (7.9). The first two proofs are for $n = 2$, i.e., $|ax + by| \leq \sqrt{(a^2 + b^2)} \times \sqrt{(x^2 + y^2)}$, and employ non-isometric area transformations. Since $|ax + by| \leq |a||x| + |b||y|$, it suffices to show $|a||x| + |b||y| \leq \sqrt{(a^2 + b^2)}\sqrt{(x^2 + y^2)}$. The third proof is valid when $n \geq 2$.

Proof 1 [Nelsen, 1994a]. Since the area of a parallelogram with given sides is less than or equal to the area of a rectangle with the same given sides, the area of the rectangle in Figure 7.8a (the four shaded triangles and the unshaded parallelogram) is less than or equal to the area of the polygon (the four triangles and the unshaded rectangle) in Figure 7.8b. Hence

$$(|a| + |y|)(|b| + |x|) \leq 2\left(\frac{1}{2}|a||b| + \frac{1}{2}|x||y|\right) + \sqrt{a^2 + b^2}\sqrt{x^2 + y^2},$$

so

$$|a||x| + |b||y| \leq \sqrt{a^2 + b^2}\sqrt{x^2 + y^2}. \tag{7.11}$$

Proof 2 [Alsina, 2004a]. In this proof (see Figure 7.9), we construct rectangles with dimensions $|a| \times |x|$ and $|b| \times |y|$ on the legs of a right triangle,

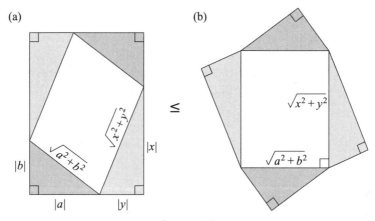

Figure 7.8.

and transform them into parallelograms with the same area. In Figure 7.9d, we note that, given a parallelogram and rectangle with the same side lengths, the rectangle has the greater area. Hence

$$|a|\,|x| + |b|\,|y| \le \sqrt{a^2 + b^2}\,\sqrt{x^2 + y^2}.$$

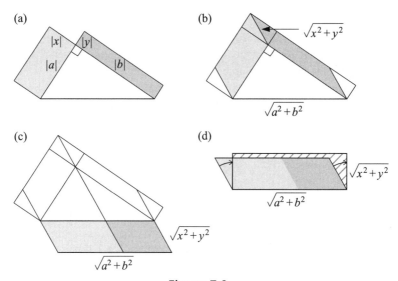

Figure 7.9.

The third proof uses vector notation. Let $\mathbf{a} = (a_1, a_2, \ldots, a_n)$ and $\mathbf{b} = (b_1, b_2, \ldots, b_n)$ be two vectors. The *dot product* $\mathbf{a} \cdot \mathbf{b}$ is the scalar (real number) defined by $\mathbf{a} \cdot \mathbf{b} = a_1 b_1 + a_2 b_2 + \cdots + a_n b_n$; while the *length* $\|\mathbf{a}\|$ of a vector $\mathbf{a}$ is the scalar defined by $\|\mathbf{a}\| = \sqrt{\mathbf{a} \cdot \mathbf{a}} = \sqrt{a_1^2 + a_2^2 + \cdots + a_n^2}$. With vector notation, the Cauchy-Schwarz inequality can be expressed succinctly as $|\mathbf{a} \cdot \mathbf{b}| \le \|\mathbf{a}\| \, \|\mathbf{b}\|$. This proof uses vectors and a projection, along with the Pythagorean comparison from Chapter 1.

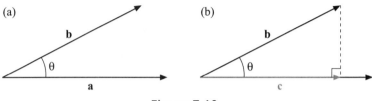

Figure 7.10.

Proof 3. In Figure 7.10a we see a position representation of two vectors $\mathbf{a}$ and $\mathbf{b}$ with a common initial point. Let θ denote the angle between $\mathbf{a}$ and $\mathbf{b}$, and let $\mathbf{c}$ be the vector projection of $\mathbf{b}$ on $\mathbf{a}$, as illustrated in Figure 7.10b. Then, $\|\mathbf{c}\| = \|\mathbf{b}\| \cos \theta = |\mathbf{a} \cdot \mathbf{b}|/\|\mathbf{a}\|$. But $\|\mathbf{c}\| \le \|\mathbf{b}\|$, and hence $|\mathbf{a} \cdot \mathbf{b}| \le \|\mathbf{a}\| \, \|\mathbf{b}\|$, as claimed.

Proof 4. We conclude this section with another algebraic proof, showing that the Cauchy-Schwarz inequality for n numbers is a consequence of the AM-GM inequality for two numbers. Given, as above, $\mathbf{a} = (a_1, a_2, \ldots, a_n)$ and $\mathbf{b} = (b_1, b_2, \ldots, b_n)$, we consider the two vectors $\mathbf{a}/\|\mathbf{a}\|$ and $\mathbf{b}/\|\mathbf{b}\|$. These vectors each have length 1. For each i between 1 and n, the AM-GM inequality yields

$$\frac{|a_i|}{\|\mathbf{a}\|} \cdot \frac{|b_i|}{\|\mathbf{b}\|} \le \frac{(a_i/\|\mathbf{a}\|)^2 + (b_i/\|\mathbf{b}\|)^2}{2}.$$

Summing on i from 1 to n gives

$$\sum_{i=1}^{n} \frac{|a_i|}{\|\mathbf{a}\|} \cdot \frac{|b_i|}{\|\mathbf{b}\|} \le \frac{1}{2} \left(\sum_{i=1}^{n} \frac{a_i^2}{\|\mathbf{a}\|^2} + \sum_{i=1}^{n} \frac{b_i^2}{\|\mathbf{b}\|^2} \right) = 1,$$

and hence

$$|\mathbf{a} \cdot \mathbf{b}| \le |a_1| \, |b_1| + |a_2| \, |b_2| + \cdots + |a_n| \, |b_n| \le \|\mathbf{a}\| \, \|\mathbf{b}\|.$$

Application 7.3. *Extrema of a linear function on an ellipsoid*

Let $a, b, c, p, q,$ and r be real numbers, with $p, q, r > 0$. What are the largest and smallest values of $f(x, y, z) = ax + by + cz$ for points (x, y, z) on the ellipsoid $x^2/p + y^2/q + z^2/r = 1$? From the Cauchy-Schwarz inequality, we have

$$[f(x, y, z)]^2 = (ax + by + cz)^2$$
$$= \left(a\sqrt{p} \cdot \frac{x}{\sqrt{p}} + b\sqrt{q} \cdot \frac{y}{\sqrt{q}} + c\sqrt{r} \cdot \frac{z}{\sqrt{r}} \right)^2$$
$$\leq (a^2 p + b^2 q + c^2 r) \left(\frac{x^2}{p} + \frac{y^2}{q} + \frac{z^2}{r} \right)$$
$$= a^2 p + b^2 q + c^2 r$$

with equality if and only if $x/ap = y/bq = z/cr$. Consequently, the extreme values of $f(x, y, z)$ are $\sqrt{a^2 p + b^2 q + c^2 r}$ and $-\sqrt{a^2 p + b^2 q + c^2 r}$, and they occur, respectively, at the points $(ap/M, bq/M, cr/M)$ and $(-ap/M, -bq/M, -cr/M)$ where $M = \sqrt{a^2 p + b^2 q + c^2 r}$.

Application 7.4. *The triangle inequality for vectors*

In Chapter 1 we proved a special case of Minkowski's inequality: for positive a_i and b_i, we have

$$\sqrt{\left(\sum_{i=1}^{n} a_i \right)^2 + \left(\sum_{i=1}^{n} b_i \right)^2} \leq \sum_{i=1}^{n} \sqrt{a_i^2 + b_i^2}.$$

We now prove this closely related inequality: for real a_i and b_i, we have

$$\sqrt{\sum_{i=1}^{n} a_i^2} + \sqrt{\sum_{i=1}^{n} b_i^2} \geq \sqrt{\sum_{i=1}^{n} (a_i + b_i)^2},$$

i.e., the triangle inequality $\|\mathbf{a}\| + \|\mathbf{b}\| \geq \|\mathbf{a} + \mathbf{b}\|$ for the vectors $\mathbf{a} = (a_1, a_2, \dots, a_n)$ and $\mathbf{b} = (b_1, b_2, \dots, b_n)$.

First, note that

$$\sum_{i=1}^{n} (a_i + b_i)^2 = \left| \sum_{i=1}^{n} a_i (a_i + b_i) + \sum_{i=1}^{n} b_i (a_i + b_i) \right|$$
$$\leq \left| \sum_{i=1}^{n} a_i (a_i + b_i) \right| + \left| \sum_{i=1}^{n} b_i (a_i + b_i) \right|,$$

and hence from the Cauchy-Schwarz inequality, we have

$$\left| \sum_{i=1}^{n} a_i (a_i + b_i) \right| \leq \sqrt{\sum_{i=1}^{n} a_i^2} \sqrt{\sum_{i=1}^{n} (a_i + b_i)^2}$$

and

$$\left| \sum_{i=1}^{n} b_i \left(a_i + b_i\right) \right| \leq \sqrt{\sum_{i=1}^{n} b_i^2} \sqrt{\sum_{i=1}^{n} \left(a_i + b_i\right)^2},$$

so

$$\sum_{i=1}^{n} \left(a_i + b_i\right)^2 \leq \left(\sqrt{\sum_{i=1}^{n} a_i^2} + \sqrt{\sum_{i=1}^{n} b_i^2} \right) \sqrt{\sum_{i=1}^{n} \left(a_i + b_i\right)^2},$$

which, upon division of both sides by $\sqrt{\sum_{i=1}^{n} \left(a_i + b_i\right)^2}$, yields the desired result.

Application 7.5. *The sample correlation coefficient*

In statistics, the sample correlation coefficient r_{xy} is a measure of the strength and direction of a linear relationship between the x- and y-values in a set of n ordered pairs $\{(x_1, y_1), (x_2, y_2), \dots, (x_n, y_n)\}$. Letting $\bar{x} = (x_1 + x_2 + \cdots + x_n)/n$ and $\bar{y} = (y_1 + y_2 + \cdots + y_n)/n$, we have

$$r_{xy} = \frac{\sum_{i=1}^{n} (x_i - \bar{x})(y_i - \bar{y})}{\sqrt{\sum_{i=1}^{n} (x_i - \bar{x})^2} \sqrt{\sum_{i=1}^{n} (y_i - \bar{y})^2}}.$$

If we set $a_i = x_i - \bar{x}$ and $b_i = y_i - \bar{y}$, then r_{xy} can be expressed more compactly in terms of vector notation: $r_{xy} = \mathbf{a} \cdot \mathbf{b} / \|\mathbf{a}\| \|\mathbf{b}\|$. As a consequence of the Cauchy-Schwarz inequality, $|r_{xy}| \leq 1$ with equality if and only if the x- and y-values are proportional, or equivalently, if they all lie on a line.

Inequalities and names

Since there is ample evidence that Bunyakovski's work preceded that of Schwarz, calling (7.9) the *Cauchy-Schwarz* inequality may seem inappropriate, even unjust. However, by modern standards, both Bunyakovski and Schwarz should consider themselves fortunate to have their names associated with such an important result. It is rare today to receive much credit for merely finding a continuous analog of a discrete inequality (or vice versa).

The names of many inequalities are merely descriptive (e.g., the arithmetic mean-geometric mean inequality); others are named for a mathematician. However, there is no rule governing that naming—sometimes it is for the one who first discovered the inequality, but sometimes it is for the one who crafted its final form or the one responsible for an important application [Steele, 2004].

7.4 Aczél's inequality

A celebrated inequality due to János Aczél [Bullen, 1998] states that if $a_1, a_2, \ldots, a_n$ and $b_1, b_2, \ldots, b_n$ are two sets of real numbers such that $a_1^2 - a_2^2 - a_3^2 - \cdots - a_n^2 > 0$ and $b_1^2 - b_2^2 - b_3^2 - \cdots - b_n^2 > 0$, then

$$\sqrt{a_1^2 - a_2^2 - \cdots - a_n^2}\sqrt{b_1^2 - b_2^2 - \cdots - b_n^2} \le |a_1 b_1 - a_2 b_2 - \cdots - a_n b_n|,$$
(7.12)

with equality if and only if the components of the vectors are proportional. We will use a visual argument to establish the $n = 2$ case, and then proceed to the general case.

In Figure 7.8a, let $|c| = \sqrt{a^2 + b^2}$ and $|z| = \sqrt{x^2 + y^2}$. Then $|a| = \sqrt{c^2 - b^2}$ and $|x| = \sqrt{z^2 - y^2}$, so (7.11) becomes $\sqrt{c^2 - b^2}\sqrt{z^2 - y^2} + |b||y| \le |c||z|$, or

$$\sqrt{c^2 - b^2}\sqrt{z^2 - y^2} \le |c||z| - |b||y| \le |cz - by|,$$

which is the $n = 2$ case of (7.12) with $c = a_1, b = a_2, z = b_1$, and $y = b_2$.

To establish (7.12) in general, let $c = a_1$, $b^2 = a_2^2 + a_3^2 + \cdots + a_n^2$, $z = b_1$, and $y^2 = b_2^2 + b_3^2 + \cdots + b_n^2$. Then

$$\sqrt{a_1^2 - a_2^2 - \cdots - a_n^2}\sqrt{b_1^2 - b_2^2 - \cdots - b_n^2} + |a_2 b_2 + a_3 b_3 + \cdots + a_n b_n|$$
$$\le \sqrt{c^2 - b^2}\sqrt{z^2 - y^2} + |b||y|$$
$$\le |c||z| = |a_1||b_1|,$$

and the result follows.

János Aczél

Figure 7.11.

Application 7.6. *The Neuberg-Pedoe inequality.*

In 1941, while doing some mathematics during an air raid on Southampton, the British geometer Dan Pedoe (1910–1998) discovered the following remarkable two-triangle inequality: given two triangles with sides of length a_i, b_i, c_i and area $K_i, i = 1, 2$, we have

$$a_1^2(-a_2^2 + b_2^2 + c_2^2) + b_1^2(a_2^2 - b_2^2 + c_2^2) + c_1^2(a_2^2 + b_2^2 - c_2^2) \geq 16K_1K_2,$$

with equality if and only if the triangles are similar [Pedoe, 1963, 1970, 1991]. Later Pedoe learned that Jean Baptiste Joseph Neuberg (1840–1926) had discovered the inequality in the 19th century, but had not proved that equality implies the similarity of the triangles.

An inequality of this sort is unusual in the sense that it involves two unrelated triangles. As noted by L. Carlitz [Carlitz, 1971], the Neuberg-Pedoe inequality is a consequence of Aczél's inequality (7.12) for $n = 4$, with (a_1, a_2, a_3, a_4) replaced by $(a_1^2 + b_1^2 + c_1^2, \sqrt{2}a_1^2, \sqrt{2}b_1^2, \sqrt{2}c_1^2)$ with (b_1, b_2, b_3, b_4) replaced by $(a_2^2 + b_2^2 + c_2^2, \sqrt{2}a_2^2, \sqrt{2}b_2^2, \sqrt{2}c_2^2)$, and Heron's formula (see Application 4.1) where $16K_i^2 = (a_i^2 + b_i^2 + c_i^2)^2 - 2(a_i^4 + b_i^4 + c_i^4)$.

7.5 Challenges

7.1 In triangle ABC, show that (a) $\csc(A/2) + \csc(B/2) + \csc(C/2) \geq 6$ and (b) $s \leq 3R\sqrt{3}/2$, where s denotes the semiperimeter and R the circumradius.

7.2 Create a visual proof to show that for any real a, b, θ,

$$-\sqrt{a^2 + b^2} \leq a\cos\theta + b\sin\theta \leq \sqrt{a^2 + b^2}.$$

7.3 Prove that the Cauchy-Schwarz inequality implies the arithmetic mean-root mean square inequality.

7.4 (a) Prove *Brahmagupta's* identity: for real numbers a, b, t, x, y,

$$(a^2 + tb^2)(x^2 + ty^2) = (ax + tby)^2 + t(ay - bx)^2.$$

(b) Show that Brahmagupta's identity implies Diophantus' identity, and hence the $n = 2$ case of the Cauchy-Schwarz inequality;

(c) Show that Brahmagupta's identity implies the $n = 2$ case of Aczél's inequality.

7.5 Let a, b, c be real numbers such that $a^2 + b^2 + c^2 = 1$. What is the minimum value of $ab + bc + ca$?

First solution: By the Cauchy-Schwarz inequality,

$$|ab + bc + ca| \leq \sqrt{a^2 + b^2 + c^2} \sqrt{b^2 + c^2 + a^2} = 1,$$

so $ab + bc + ca \geq -1$.

Second solution: Since $(a + b + c)^2 \geq 0$, $1 + 2(ab + bc + ca) \geq 0$, and hence $ab + bc + ca \geq -1/2$.

Which solution (if either) is correct? [Barbeau, 2000]

7.6 If ABC is an acute triangle, prove that $m_a + m_b + m_c \leq 4R + r$. (Hint: Use Carnot's Theorem 7.3.)

7.7 Given a triangle ABC, find the square with maximal area inscribed in the triangle with one side on BC.

7.8 If A, B, and C are the angles of an acute triangle, show that $\cos^2 A + \cos^2 B + \cos^2 C < 1$.

7.9 Let P be a rectangular box with sides a, b, c; diagonal of length $d = \sqrt{a^2 + b^2 + c^2}$, and total lateral area $L = 2ac + 2bc$, show *Gotman's inequality* $L \leq \sqrt{2}d^2$ [Mitrinović et al., 1989].

7.10 In the fourth proof of the Cauchy-Schwarz inequality (see Section 7.3) we showed that the AM-GM inequality for two numbers implies the Cauchy-Schwarz inequality for n numbers. Prove that the two inequalities are actually equivalent by proving the converse.

7.11 What inequality is illustrated in Figure 7.12 [Kung, 2008]?

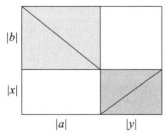 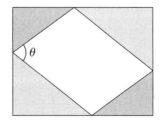

Figure 7.12.

7.12 In $\triangle ABC$, let u, v, w denote the lengths of the perpendiculars from the circumcenter to the sides, and R the circumradius (see Figure 6.11). In Corollary 6.1, we showed that $u + v + w > R$. Prove that we also have an upper bound: $u + v + w \leq 3R/2$.

CHAPTER 8

Employing graphs of functions

Many simple properties of real-valued functions, such as boundedness, monotonicity, convexity, and the Lipschitz condition, can be expressed in terms of inequalities. Consequently there are visual representations of many of them, some of which are familiar. In this chapter we introduce the idea of a *moving frame* to illustrate some of these properties, and then use them to establish additional inequalities. We also investigate the role played by the convexity or concavity of a function in establishing functional inequalities. We conclude the chapter by examining inequalities in which areas under graphs of functions represent numbers.

8.1   Boundedness and monotonicity

Let S and T be subsets of the reals. A function $f : S \to T$ is *bounded* if there exist constants m and M such that for all x in S, $m \le f(x) \le M$. Visually, this means that the entire graph of $y = f(x)$ lies between the horizontal lines $y = m$ and $y = M$, as illustrated in Figure 8.1a. A *moving frame* is like a window, in this case with height $M - m$ units and some convenient width, such that, as the frame is moved horizontally with the opening always between the lines $y = m$ and $y = M$, we see the graph inside the frame, never in the opaque regions shaded gray in Figure 8.1b.

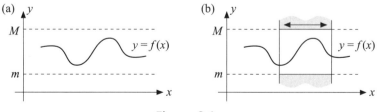

Figure 8.1.

A function $f : S \rightarrow T$ is *nondecreasing* if f satisfies the inequality $f(a) \leq f(b)$ for all $a \leq b$ in S, and analogously, f is *nonincreasing* if f satisfies $f(a) \geq f(b)$ for all $x \leq y$ in S. Whenever one of these conditions holds throughout S, we say that f is *monotonic* or *monotone*. A moving frame to show that a given function f is, say, nondecreasing, consists of a window formed by a pair of axes in which the second and fourth quadrants are opaque, and which moves along the curve with its origin on the graph of the function. The function is nondecreasing if one always sees the graph in the frame, here the first and third quadrants, as illustrated in Figure 8.2.

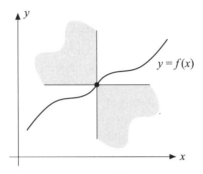

Figure 8.2.

Later in this chapter we will encounter a similar frame, transparent between lines parallel to $y = \pm Mx$, for illustrating the Lipschitz condition.

Application 8.1. *An inequality of L. Bankoff* [Mitrinović et al., 1989]

If r and R are, respectively, the inradius and circumradius of a triangle, then

$$\frac{R}{r} + \frac{r}{R} \geq \frac{5}{2}.$$

Letting $x = R/r$, we need to show that $x + (1/x) \geq 5/2$, or $x^2 - (5/2)x + 1 \geq 0$ when $x \geq 2$ (recall $R \geq 2r$ by Euler's inequality in Section 4.1). But $x^2 - (5/2)x + 1 = (x - 1/2)(x - 2)$, so the graph of $y = x^2 - (5/2)x + 1$ is nonnegative (and increasing) for $x \geq 2$.

Application 8.2. *Edges and volume in a class of prisms*

For any right regular prism (a solid with two identical regular n-gons as bases and n identical rectangles as lateral faces), if V denotes the volume and E the total edge length, then $E^3/V \geq 972\sqrt{3}$. To prove this surprising inequality, we first note that $E = nh + 2an$ and $V = (nha^2/4) \cot(\pi/n)$, where a denotes the length of each side of the n-gon, and h the height of the prism. Hence we have

$$\frac{E^3}{V} = \frac{n^3(h+2a)^3}{(nha^2/4)\cot(\pi/n)} = 4n^2\tan(\pi/n)\frac{(h+a+a)^3}{h \cdot a \cdot a}$$

$$\geq 108n^2\tan(\pi/n),$$

where in the last step we used the AM-GM inequality. Now let $f(x) = x^2\tan(\pi/x)$. Note that $f(3) = 9\sqrt{3} < 16 = f(4)$. We claim that f is increasing for $x \geq 4$, so the minimum value of $n^2\tan(\pi/n)$ for integers $n \geq 3$ is $9\sqrt{3}$. We have

$$f'(x) = \frac{2x\sin(\pi/x)\cos(\pi/x) - \pi}{\cos^2(\pi/x)},$$

so $f'(x) > 0$ if $2x\sin(2\pi/x) > \pi$, or if $\sin(2\pi/x) > \pi/(2x)$ for $x \geq 4$. But $2\pi/x \in (0, \pi/2]$, so by Jordan's inequality (see Section 1.7),

$$\sin\frac{2\pi}{x} \geq \frac{2}{\pi} \cdot \frac{2\pi}{x} = \frac{4}{x} > \frac{\pi}{2x}.$$

Hence the minimum value of $n^2\tan(\pi/n)$ is $9\sqrt{3}$, so $E^3/V \geq 972\sqrt{3}$ as claimed. We have equality for triangular prisms with edges all the same length.

Application 8.3. *The Cauchy-Schwarz inequality revisited*
  In Section 7.3 we gave four proofs of the Cauchy-Schwarz inequality

$$\left(\sum_{i=1}^{n} a_i b_i\right)^2 \leq \left(\sum_{i=1}^{n} a_i^2\right)\left(\sum_{i=1}^{n} b_i^2\right),$$

two geometric proofs valid for $n = 2$, and vector and algebraic proofs for $n \geq 2$. We now give a functional proof for $n \geq 2$. Define a function f with domain $(-\infty, \infty)$ by

$$f(x) = \sum_{i=1}^{n} (a_i x + b_i)^2$$

$$= \left(\sum_{i=1}^{n} a_i^2\right)x^2 + 2\left(\sum_{i=1}^{n} a_i b_i\right)x + \left(\sum_{i=1}^{n} b_i^2\right).$$

  The graph of $y = f(x)$ is a parabola opening upwards, and since f is a sum of squares, $f(x) \geq 0$ on $(-\infty, \infty)$. Thus the discriminant of the quadratic is less than or equal to zero, so

$$4\left(\sum_{i=1}^{n} a_i b_i\right)^2 - 4\left(\sum_{i=1}^{n} a_i^2\right)\left(\sum_{i=1}^{n} b_i^2\right) \leq 0,$$

from which the Cauchy-Schwarz inequality follows.

Application 8.4.   *The statistical regression line*

In Application 7.5, we examined the correlation coefficient, a measure of the strength and direction of linear relationship between the x- and y-values in a set of n ordered pairs $\{(x_1, y_1), (x_2, y_2), \ldots, (x_n, y_n)\}$. We now find the line $y = mx + c$ which best fits this set of ordered pairs, in the sense that the sum of squared distances $[y_i - (mx_i + c)]^2$ is a minimum. This line is called the *least squares line*, or the *regression line of y on x*.

Our task is to find the numbers m and c (in terms of the given x- and y-values) that minimize

$$Q(m, c) = \sum_{i=1}^{n} [y_i - (mx_i + c)]^2.$$

It will be convenient to write, as we did in Application 7.5, $a_i = x_i - \bar{x}$ and $b_i = y_i - \bar{y}$, where $\bar{x}$ and $\bar{y}$ denote the (arithmetic) means of the given x- and y-values. Because $\sum_{i=1}^{n} a_i = \sum_{i=1}^{n} b_i = 0$,

$$\begin{aligned}
Q(m, c) &= \sum_{i=1}^{n} [(b_i - ma_i) + (\bar{y} - m\bar{x} - c)]^2 \\
&= \sum_{i=1}^{n} (b_i - ma_i)^2 + 2(\bar{y} - m\bar{x} - c) \sum_{i=1}^{n} (b_i - ma_i) \\
&\quad + n(\bar{y} - m\bar{x} - c)^2 \\
&= \sum_{i=1}^{n} (b_i - ma_i)^2 + n(\bar{y} - m\bar{x} - c)^2.
\end{aligned}$$

To minimize $Q(m, c)$ we will find the value of m which minimizes the first term $q(m) = \sum_{i=1}^{n} (b_i - ma_i)^2$ in $Q(m, c)$, and then set $c = \bar{y} - m\bar{x}$ so $n(\bar{y} - m\bar{x} - c)^2 = 0$. Now

$$\begin{aligned}
q(m) &= \sum_{i=1}^{n} (b_i - ma_i)^2 = \sum_{i=1}^{n} b_i^2 - 2m \sum_{i=1}^{n} a_i b_i + m^2 \sum_{i=1}^{n} a_i^2 \\
&= Am^2 + Bm + C,
\end{aligned}$$

where $A = \sum_{i=1}^{n} a_i^2$, $B = -2 \sum_{i=1}^{n} a_i b_i$, and $C = \sum_{i=1}^{n} b_i^2$. Since the graph of q is a parabola opening upwards, its minimum value occurs at $m = -B/(2A)$, and thus the slope and intercept of the least squares line are

$$m = \frac{\sum_{i=1}^{n} a_i b_i}{\sum_{i=1}^{n} a_i^2} = \frac{\mathbf{a} \cdot \mathbf{b}}{\|\mathbf{a}\|^2} = \frac{\sum_{i=1}^{n} (x_i - \bar{x})(y_i - \bar{y})}{\sum_{i=1}^{n} (x_i - \bar{x})^2} \quad \text{and} \quad c = \bar{y} - m\bar{x}.$$

In terms of the correlation coefficient r_{xy} from Application 7.5, $m = r_{xy} \|\mathbf{b}\| / \|\mathbf{a}\|$, and consequently $|m| \leq \|\mathbf{b}\| / \|\mathbf{a}\|$.

Application 8.5. *Inequalities in convex polyhedra*

In 1752, Leonhard Euler (1707–1783) discovered the remarkable formula $F + V - E = 2$, where F, V, and E are, respectively, the number of faces, vertices, and edges in a convex polyhedron. Euler and this formula were honored in 2007 when Switzerland, the country of Euler's birth, issued a postage stamp commemorating the 300th anniversary of that event. See Figure 8.3.

Figure 8.3.

If a denotes the average number of edges per vertex, then $a = 2E/V$ (since each edge connects two vertices), and $a \geq 3$ (since at least three edges meet at each vertex), so $2E \geq 3V$. Similarly, if b denotes the average number of edges per face, then $b = 2E/F$ (since each edge borders two faces), and $b \geq 3$ (since each face has at least three edges), so $2E \geq 3F$.

From these simple observations, many additional relationships (both equalities and inequalities) have been proved about F, V, and E [Euler, 1767; Féjes Toth, 1948, 1958; Catalan, 1865; etc.] using algebraic manipulations. In this section we will derive a few visually.

We begin by graphing $E = V + F - 2$, where the number F of faces is fixed (see Figure 8.4). Since $E \geq 3V/2$ and $E \geq 3F/2$, only the points (V, E) on the line $E = V + F - 2$ above the lines $E = 3V/2$ and $E = 3F/2$ are permissible, i.e., the points on the dark segment of the line $E = V + F - 2$. Since the endpoints of the segment are $(2 + F/2, 3F/2)$ and $(2F - 4, 3F - 6)$, it follows that $2 + F/2 \leq V \leq 2F + 4$ and $E \leq 3F - 6$. Thus we have proved

Theorem 8.1 (Euler-Steinitz-Féjes Toth). *$E \leq 3F - 6$, $V \leq 2F + 4$, and $V \geq 2 + F/2$.*

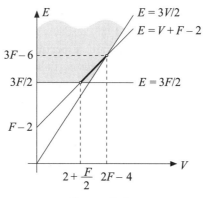

Figure 8.4.

The latest medical obsession: body mass index

Recently much publicity has been given to the so-called *body mass index BMI* $= w/h^2$, where w is a person's weight in kilograms and h the height in meters. For adults, h is fixed, so that BMI is proportional to weight (which for many of us is not constant). According to [Saldaña, 1994], normal weight corresponds approximately to the inequality $20 \le BMI \le 25$, or $20h^2 \le w \le 25h^2$.

Before *BMI* was introduced, a popular European recommendation [Alsina, 2004b] for w in relation to h was the approximation $w \approx 100h - 100$, i.e., a weight in kilograms near the number of centimeters of height exceeding one meter (e.g., 85 kilograms for a person 185 centimeters tall). If we follow this rule, do we need to check the inequality $20 \le BMI \le 25$?

To satisfy $20 \le (100h - 100)/h^2 \le 25$, we need to satisfy

$$20h^2 - 100h + 100 \le 0 \le 25h^2 - 100h + 100.$$

The second inequality is equivalent to $25(h - 2)^2 \ge 0$, which is always true. The first inequality is equivalent to $h^2 - 5h + 5 \le 0$. Examining the graph of the parabola $y = h^2 - 5h + 5$ shows that inequality holds whenever

$$1.38\text{m} \approx \frac{5 - \sqrt{5}}{2} \le h \le \frac{5 + \sqrt{5}}{2} \approx 3.61\text{m}.$$

Do you think the new index is necessary?

8.2 Continuity and uniform continuity

Inequality is the cause of all local
movements.
Leonardo da Vinci

The formal ε-δ definition of continuity of a function at a point goes as follows: A function f is continuous at a number a in its domain if for any $\varepsilon > 0$, there exists a $\delta > 0$, which may depend on both ε and a, such that

$$|f(x) - f(a)| < \varepsilon \quad \text{whenever} \quad 0 < |x - a| < \delta.$$

This definition captures the key idea of continuity by means of a relationship between *local* inequalities, local in the sense that they hold in a "window" centered at $(a, f(a))$. This can be visualized as illustrated in Figure 8.5.

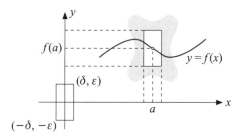

Figure 8.5.

We fix the point $(a, f(a))$, then for any $\varepsilon > 0$ we look for a $\delta > 0$ such that the rectangular window with width 2δ and height 2ε contains all the points $(x, f(x))$ for x between $a - \delta$ and $a + \delta$, so the graph enters the window along the left side and exits the window along the right side, without leaving the window at the top or bottom.

The window cannot be viewed as a moving frame, because its shape may depend on the point $(a, f(a))$, i.e., for the same (arbitrary) $\varepsilon > 0$, the choice of δ is pointwise dependent. In Figure 8.6, we see that for the function $f(x) = 1/x$ and a fixed ε, the width 2δ of the frame depends on the x-coordinate, e.g., δ must be smaller when the value of x is close to 0.

In Figure 8.7 we illustrate a function that is discontinuous at a, so it is possible to construct a window that does not capture the graph of $y = f(x)$ for points near a.

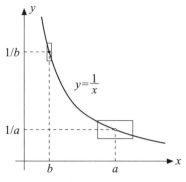

Figure 8.6.

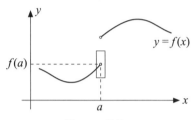

Figure 8.7.

In some cases it is possible to construct a 2δ-by-2ε window that can serve as a moving frame, that is, for any $\varepsilon > 0$, we can find a $\delta > 0$ independent of a such that the same 2δ-by-2ε window centered at $(a, f(a))$ for any a in the domain of f captures all the points on the graph for x between $a - \delta$ and $a + \delta$. Figure 8.5 illustrates this. When such a frame exists with its width depending only on the height and not the coordinates of the center of the frame, we say that f is *uniformly continuous* on its domain. Figure 8.6 illustrates a function $f(x) = 1/x$ that is not uniformly continuous on, for example, $(0, \infty)$.

8.3   The Lipschitz condition

A function f satisfies the *Lipschitz condition with constant* $M \geq 0$ [Rudolf Otto Sigismund Lipschitz (1832–1903)] if for all x_1 and x_2 in the domain of f, we have

$$|f(x_2) - f(x_1)| \leq M\,|x_2 - x_1|.$$

To visualize the Lipschitz condition with constant M [de Guzmán, 1996], we consider a pair of lines $y = Mx$ and $y = -Mx$, and translate this pair to a point $(a, f(a))$ on the graph. Then the entire graph of $y = f(x)$ lies between the lines $y = f(a) + M(x - a)$ and $y = f(a) - M(x - a)$, as seen in Figure 8.8.

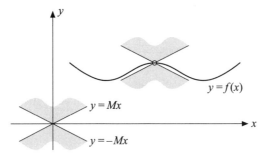

Figure 8.8.

Satisfying a Lipschitz condition is a global property of the function, and it implies continuity. If the derivative of f is bounded, then f is Lipschitz. For example, $f(x) = \sin x$ is Lipschitz with constant 1, since $|\cos x| \leq 1$. However, there are very smooth functions which do not satisfy a Lipschitz condition on the domain for any $M \geq 0$, for example, $f(x) = e^x$.

8.4 Subadditivity and superadditivity

A function $f : [0, \infty) \to [0, \infty)$ is *subadditive* if for all nonnegative a and b, we have

$$f(a + b) \leq f(a) + f(b).$$

An example is the square root function $f(x) = \sqrt{x}$, as illustrated in Figure 1.3. We can illustrate the subadditivity of $f(x)$ visually by comparing the graph of $y = f(x)$ to the graph of $y = f(a) + f(x - a)$, where a is an arbitrary positive number. This graph is the same as the graph of f, but translated a units to the right and $f(a)$ up, as illustrated in Figure 8.9. When f is subadditive, the graph of $y = f(x)$ will lie below the graph of $y = f(a) + f(x-a)$, so when $x = a+b$, we have $f(a+b) \leq f(a) + f(b)$ [de Guzmán, 1996].

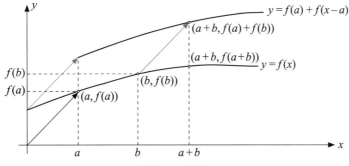

Figure 8.9.

It is easy to show that all nonincreasing functions are subadditive, as are all linear functions $f(x) = mx + c$ with $m, c \geq 0$. The illustration in Figure 8.9 is also useful for showing that many functions are not subadditive, e.g. $f(x) = e^x$.

Superadditivity is the dual condition of subadditivity, i.e., a function f : $[0, \infty) \to [0, \infty)$ is *superadditive* if for all nonnegative a and b, we have

$$f(a + b) \geq f(a) + f(b).$$

In Challenge 8.16 you will investigate the geometric behavior of superadditive functions. All superadditive functions are nondecreasing, and for a one-to-one function f, the superadditivity of f is equivalent to the subadditivity of f^{-1}. Functions that are both subadditive and superadditive are called *additive*.

Application 8.6. *Transforming the sides of a triangle*

Suppose a, b, and c are positive numbers that are the sides of a triangle (i.e., $a \leq b + c$, etc.). What properties of a function f insure that $f(a)$, $f(b)$, and $f(c)$ also form a triangle? One answer is *nondecreasing subadditive* functions. If f : $[0, \infty) \to [0, \infty)$ is nondecreasing and subadditive, then $0 \leq f(a) \leq f(b + c) \leq f(b) + f(c)$ (and similarly for $b \leq a + c$ and $c \leq a + b$), so $f(a)$, $f(b)$, and $f(c)$ form a triangle. For example, $\sqrt{a}$, $\sqrt{b}$, and $\sqrt{c}$ form a triangle ($f(x) = \sqrt{x}$), as do $a/(a + 1)$, $b/(b + 1)$, and $c/(c + 1)$ ($f(x) = x/(x + 1)$).

8.5   Convexity and concavity

A set C of points in the plane is *convex* if for any pair P, Q of points in C, the line segment joining P and Q belongs to C. A function f whose domain is an interval I is convex if the set of points above its graph is convex, i.e., if the set $C = \{(x, y) \mid x \in I, y \geq f(x)\}$ is convex. Thus a convex function satisfies the inequality

$$f(ta + (1 - t)b) \leq tf(a) + (1 - t)f(b) \tag{8.1}$$

for all a, b in I and all t such that $0 < t < 1$. The inequality follows because the line segment connecting the points $(a, f(a))$ and $(a, f(a))$ must lie on or above the graph of $y = f(x)$ when f is convex, as illustrated in Figure 8.10.

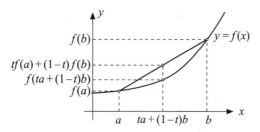

Figure 8.10.

Thus the convexity of f implies that the chord connecting any two points on the graph of $y = f(x)$ lies above that graph. If the inequality in (8.1) is strict, then we say that f is *strictly convex* on I.

We define concavity from convexity: A function f whose domain is an interval I is *concave* if $-f$ is convex; or if the set $C = \{(x, y)| x \in I, y \leq f(x)\}$ is convex. Thus for concave functions, the inequality in (8.1) is reversed, and a chord connecting any two points on the graph of $y = f(x)$ lies below the graph. When the function is twice differentiable, convexity or concavity is easily established by examining the sign of the second derivative.

Application 8.7. *A property of concave functions.* Let f be a concave function defined on $[0, b)$, where b can be finite or ∞, with $f(0) = 0$. Then for any λ in $(0, 1)$ and $x \geq 0$, $\lambda f(x) \leq f(\lambda x)$. See Figure 8.11.

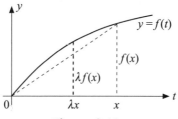

Figure 8.11.

Because $f(x) = \ln(1 + x)$ is concave on $(0, \infty)$ with $f(0) = 0$, for λ in $(0, 1)$, $\lambda \ln(1 + x) \leq \ln(1 + \lambda x)$, or $(1 + x)^{\lambda} \leq 1 + \lambda x$, which is equivalent to Bernoulli's inequality (see Challenge 1.5).

Application 8.8. *The Kantorovich inequality*

Let $0 < x_1 \leq x_2 \leq \cdots \leq x_n$ be given positive numbers, and let λ_1, $\lambda_2, \ldots, \lambda_n \geq 0$, $\lambda_1 + \lambda_2 + \cdots + \lambda_n = 1$. The Kantorovich inequality

[Leonid Vitaliyevich Kantorovich (1912–1986)] states that

$$\left(\sum_{i=1}^{n} \lambda_i x_i\right)\left(\sum_{i=1}^{n} \frac{\lambda_i}{x_i}\right) \le \frac{A^2}{G^2},$$

where A and G are, respectively, the arithmetic and geometric means of x_1 and x_n, the two extremes of the set of x values. Our proof uses the convexity of $f(x) = 1/x$.

Proof. Let $\bar{x} = \sum_{i=1}^{n} \lambda_i x_i$, and note that $\bar{x} \in [x_1, x_n]$. Let $f(x) = 1/x$, and let $g(x)$ be the linear function through the points $(x_1, 1/x_1)$ and $(x_n, 1/x_n)$, i.e.,

$$g(x) = \frac{1}{x_n} + \frac{1/x_n - 1/x_1}{x_n - x_1}(x - x_n) = \frac{x_1 + x_n - x}{x_1 x_n}.$$

Since f is convex, $f(x) \le g(x)$ on $[x_1, x_n]$, and hence

$$\sum_{i=1}^{n} \frac{\lambda_i}{x_i} = \sum_{i=1}^{n} \lambda_i f(x_i) \le \sum_{i=1}^{n} \lambda_i g(x_i) = g\left(\sum_{i=1}^{n} \lambda_i x_i\right) = g(\bar{x}).$$

Thus

$$\left(\sum_{i=1}^{n} \lambda_i x_i\right)\left(\sum_{i=1}^{n} \frac{\lambda_i}{x_i}\right) \le \bar{x} g(\bar{x}) \le \max_{x_1 \le x \le x_n} x g(x)$$

$$= \max_{x_1 \le x \le x_n} \frac{x(x_1 + x_n - x)}{x_1 x_n}.$$

Because $x(x_1 + x_n - x)$ attains its maximum value at $x = (x_1 + x_n)/2$,

$$\left(\sum_{i=1}^{n} \lambda_i x_i\right)\left(\sum_{i=1}^{n} \frac{\lambda_i}{x_i}\right) \le \frac{(x_1 + x_n)^2}{4 x_1 x_n}$$

as claimed.

When the inequality in (8.1) is satisfied for $t = 1/2$, we have a weaker inequality: for all a and b in I, $f((a+b)/2) \le [f(a) + f(b)]/2$. In this case we say that f is *midconvex* on I. It can be shown [Roberts and Varberg, 1973] that continuous midconvex functions are convex, and so if f is continuous, verifying midconvexity suffices to show convexity.

The midconvexity inequality in the preceding paragraph can be extended to three or more numbers. For example, if f is convex, then

$$f\left(\frac{a+b+c}{3}\right) = f\left(\frac{2}{3} \cdot \frac{a+b}{2} + \frac{1}{3}c\right) \le \frac{2}{3} f\left(\frac{a+b}{2}\right)$$

$$+ \frac{1}{3} f(c) \le \frac{f(a) + f(b) + f(c)}{3}.$$

A similar result (with the inequality reversed) holds for concave functions; and both can be extended to n numbers, a result known as *Jensen's inequality* [Johan Ludwig William Valdemar Jensen (1859–1925)]: When f is convex,

$$f\left(\frac{1}{n}\sum_{i=1}^{n} x_i\right) \leq \frac{1}{n}\sum_{i=1}^{n} f(x_i) \qquad (8.2)$$

(and similarly for concave functions). When f is strictly convex, equality in (8.2) implies that $x_1 = x_2 = \cdots = x_n$.

The relationship between concave (convex) functions and subadditivity (superadditivity) is explored in Challenge 8.1.

L.V. Kantorovich   J. L. W. V. Jensen

Figure 8.12.

Application 8.9. *Trigonometric inequalities*

Since the trigonometric functions are concave or convex on certain intervals, Jensen's inequality is useful for establishing inequalities involving trigonometric functions of the angles of a triangle. Since the sine is concave on $(0, \pi)$ and the cosine concave on $(0, \pi/2)$, applying Jensen's inequality to the angles in triangle ABC yields

$$\sin\frac{A}{2} + \sin\frac{B}{2} + \sin\frac{C}{2} \leq 3\sin\frac{A+B+C}{6} = 3\sin\frac{\pi}{6} = \frac{3}{2},$$

$$\cos\frac{A}{2} + \cos\frac{B}{2} + \cos\frac{C}{2} \leq 3\cos\frac{A+B+C}{6} = 3\cos\frac{\pi}{6} = \frac{3\sqrt{3}}{2},$$

$$\sin A + \sin B + \sin C \leq 3\sin\frac{A+B+C}{3} = 3\sin\frac{\pi}{3} = \frac{3\sqrt{3}}{2}, \qquad (8.3)$$

and if ABC is an acute or right triangle,

$$\cos A + \cos B + \cos C \le 3 \cos \frac{A + B + C}{3} = 3 \cos \frac{\pi}{3} = \frac{3}{2}.$$

In Application 7.1, we used the Erdös-Mordell inequality (rather than Jensen's inequality) to establish these inequalities.

Application 8.10. *The perimeter and circumradius of a triangle*

In triangle ABC, let a, b, c denote the sides, and R the circumradius. In Figure 6.11, the angle marked $\angle$ at the circumcenter has the same measure as $\angle A$, so $\sin A = a/2R$, or $a = 2R \sin A$. Similarly, $b = 2R \sin B$ and $c = 2R \sin C$. Using (8.3) yields

$$a + b + c = 2R \cdot (\sin A + \sin B + \sin C) \le 6R \sin \frac{\pi}{3} = 3\sqrt{3}R.$$

As a bonus, we have shown that for triangles inscribed in a circle of radius R, the perimeter is maximal when the triangle is equilateral.

Application 8.11. *The Weitzenböck inequality revisited* [Steele, 2004]

In Section 6.3, we presented a geometric proof of Weitzenböck's inequality: If a, b, c are the sides and K the area of triangle ABC, then $a^2 + b^2 + c^2 \ge 4\sqrt{3}K$. We can now give a second proof using the convexity of the cosecant function on $(0, \pi)$. We begin with the inequality in Lemma 2.1 and the familiar area formulas $K = (bc \sin A)/2 = (ca \sin B)/2 = (ab \sin C)/2$, which give

$$a^2 + b^2 + c^2 \ge ab + bc + ca = 2K(\csc A + \csc B + \csc C)$$
$$\ge 6K \csc \frac{A + B + C}{3} = 6K \csc \frac{\pi}{3} = 4\sqrt{3}K.$$

8.6   Tangent and secant lines

In Section 1.7, we introduced the idea of establishing inequalities by comparing the graphs of two functions. In this section we examine a special case where one function is convex or concave and the other is a secant line or a tangent line. Figure 8.10 illustrates the inequality between a convex function $y = f(x)$ and the secant line $y = [(b - x)f(a) + (x - a)f(b)]/(b - a)$. Before considering some examples, we present a lemma and a theorem about convex functions and their secant and tangent lines that enables comparisons between their graphs. The proofs are left as challenges.

Lemma 8.1. *Let f be convex on an interval I, and let P, Q, and R be points on the graph of $y = f(x)$ as in Figure 8.13. If m_{PQ}, m_{QR}, and m_{PR} denote the slopes of the segments PQ, QR, and PR, then*

$$m_{PQ} \leq m_{PR} \leq m_{QR}.$$

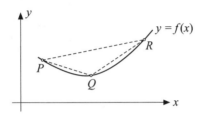

Figure 8.13.

Theorem 8.2. *Let f be differentiable and convex on an open interval I. Then*

(i) *f' is nondecreasing on I, i.e., if $a < b$ in I, then $f'(a) \leq f'(b)$;*

(ii) *for any a in I, the line tangent to f at $(a, f(a))$ lies below the graph of f, i.e., for $x \neq a$,*

$$f(x) \geq f(a) + f'(a)(x - a).$$

Analogous statements hold for concave functions, and the inequalities are strict for strictly convex and concave functions.

Example 8.1. *Bernoulli's inequality*

Bernoulli's inequality [Johann Bernoulli (1667–1748)] states that for $x > 0$ and $r > 1$, $x^r - 1 \geq r(x - 1)$. In Challenge 1.5 we established this by noting that $y = r(x-1)$ is the line tangent to the convex function $y = x^r - 1$ at $(1, 0)$.

Example 8.2. *Napier's inequality and the logarithmic mean*

Napier's inequality [John Napier (1550–1617)] establishes bounds on the difference between two logarithms: for $b > a > 0$,

$$\frac{1}{b} < \frac{\ln b - \ln a}{b - a} < \frac{1}{a}.$$

Johann Bernoulli　　　　　John Napier

Figure 8.14.

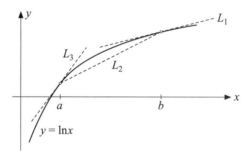

Figure 8.15.

In Figure 8.15 [Nelsen, 1993], we have a portion of the graph of the natural logarithm function, the tangent line L_1 at $(b, \ln b)$, the tangent line L_3 at $(a, \ln a)$, and the secant line L_2 joining $(a, \ln a)$ and $(b, \ln b)$. Since $y = \ln x$ is concave, the slope $1/b$ of L_1 is less than the slope $1/a$ of L_3 and by the mean value theorem, the slope of L_2 lies strictly between the slopes of L_1 and L_3 since it equals the slope of a tangent line at a point strictly between $(a, \ln a)$ and $(b, \ln b)$.

The reciprocal $(b - a)/(\ln b - \ln a)$ of the slope of the secant line L_2 in Figure 8.15 is known as the *logarithmic mean* of a and b. In Example 8.6, we will show that it lies between the arithmetic mean and geometric mean of a and b.

Example 8.3. *An exponential inequality*

In this example we show that $a^b > b^a$ whenever $e \le a < b$. Using natural logarithms, this inequality is equivalent to $\ln a/a > \ln b/b$, which holds since the function $f(x) = \ln x/x$ is decreasing for $x \ge e$. Geometrically,

$f(x)$ gives the slopes of tangent lines to the function $g(x) = (\ln x)^2/2$, and since g is concave, they are decreasing for $x \geq e$. For $a = e$ and $b = \pi$, we have $e^{\pi} > \pi^e$.

Example 8.4. *An inequality for the sides and angles in a triangle*

Let ABC be a triangle with $a > b > c$. We claim that $Ab + Bc + Ca > Ac + Ba + Cb$ [Gelca and Andreescu, 2007]. Proposition I.18 in the *Elements* of Euclid guarantees that $A > B > C$. Since the sine function is concave on $(0, \pi)$, the secant line joining $(C, \sin C)$ to $(B, \sin B)$ has a greater slope than the secant line joining $(B, \sin B)$ to $(A, \sin A)$. Thus

$$\frac{\sin B - \sin C}{B - C} > \frac{\sin A - \sin B}{A - B}.$$

Clearing fractions and simplifying yields

$$A \sin B + B \sin C + C \sin A > A \sin C + B \sin A + C \sin B.$$

If R denotes the circumradius, then as noted in Application 8.10, $a = 2R \sin A, b = 2R \sin B, c = 2R \sin C$, so multiplication by $2R$ yields the desired result.

Example 8.5. *Aristarchus' inequalities*

Aristarchus of Samos (circa 310–230 BCE) is credited by Ptolemy in *On the Size of Chords Inscribed in a Circle* with inequalities related to the length of certain chords and arcs. Using trigonometric functions, the inequalities are the following: if $0 < \beta < \alpha < \pi/2$, then

$$\frac{\sin \alpha}{\sin \beta} < \frac{\alpha}{\beta} < \frac{\tan \alpha}{\tan \beta}.$$

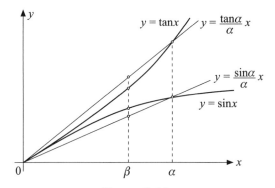

Figure 8.16.

Since the sine function is concave on $(0, \pi/2)$, the secant line $y = (\sin\alpha/\alpha)x$ joining $(0,0)$ to $(\alpha, \sin\alpha)$ lies below the graph of $y = \sin x$ for x in $(0, \alpha)$. Hence when $x = \beta$, we have $\sin\beta > (\sin\alpha/\alpha)\beta$ (see Figure 8.16), or $\sin\alpha/\sin\beta < \alpha/\beta$. Similarly, the convexity of the tangent yields $\tan\beta < (\tan\alpha/\alpha)\beta$, or $\alpha/\beta < \tan\alpha/\tan\beta$.

8.7 Using integrals

In Chapter 2 we illustrated how the representation of numbers by areas of regions in the plane can establish inequalities. In this section we continue in that vein, using integrals to compute areas.

Example 8.6. *The arithmetic mean-logarithmic mean-geometric mean inequality*

As mentioned in Example 8.2, the logarithmic mean lies between the arithmetic and geometric means: if $0 < a < b$, then

$$\sqrt{ab} < \frac{b-a}{\ln b - \ln a} < \frac{a+b}{2}.$$

To establish these inequalities, we use the fact that the area under the curve $y = 1/x$ over the interval $[a, b]$ is $\int_a^b 1/x \, dx = \ln b - \ln a$, and then approximate it using inscribed and circumscribed trapezoids.

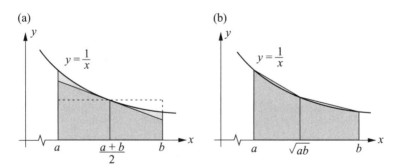

Figure 8.17.

In Figure 8.17a, the area of the inscribed trapezoid is $2(b-a)/(a+b)$, and consequently

$$\frac{2(b-a)}{a+b} < \ln b - \ln a, \text{ or } \frac{a+b}{2} > \frac{b-a}{\ln b - \ln a}.$$

In Figure 8.17b, we circumscribe two trapezoids, one with base $[a, \sqrt{ab}]$ and one with base $[\sqrt{ab}, b]$. The area of each trapezoid simplifies to $(b-a)/2\sqrt{ab}$, and hence

$$\frac{b-a}{\sqrt{ab}} > \ln b - \ln a, \text{ or } \frac{b-a}{\ln b - \ln a} > \sqrt{ab}.$$

The point $x = \sqrt{ab}$ was chosen because it minimizes the sum of the areas of the two trapezoids. The convexity of $y = 1/x$ insures that the trapezoids are inscribed and circumscribed as claimed.

Example 8.7. *The wheel of Theodorus*

The wheel of Theodorus of Cyrene (circa 5th century BCE), or the *square root spiral*, is formed from right triangles, as illustrated in Figure 8.18. Starting with an isosceles right triangle, we construct right triangles whose legs are one unit and the hypotenuse of the preceding right triangle. Thus the sequence of hypotenuses is $\sqrt{2}, \sqrt{3}, \ldots, \sqrt{n}, \ldots$ How does the spiral grow? We claim that the increase in size (as measured by the length of the hypotenuse) is proportional to half the reciprocal of the square root of n, or to be precise, for $n \geq 1$,

$$\frac{1}{2\sqrt{n+1}} < \sqrt{n+1} - \sqrt{n} < \frac{1}{2\sqrt{n}}. \tag{8.4}$$

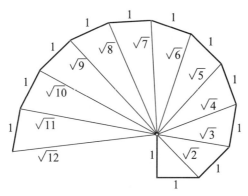

Figure 8.18.

To establish the inequalities, we note that

$$\int_n^{n+1} \frac{1}{2\sqrt{x}} \, dx = \sqrt{x}\Big|_n^{n+1} = \sqrt{n+1} - \sqrt{n},$$

and inscribe and circumscribe rectangles with areas $1/(2\sqrt{n+1})$ and $1/(2\sqrt{n})$, respectively, as illustrated in Figure 8.19.

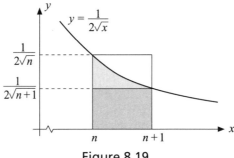

Figure 8.19.

Example 8.8. *Young's inequality*

In this section we describe an inequality due to William Henry Young
(1863–1942) [Young, 1912] that relates the values of two integrals. Let φ
and ψ be continuous strictly increasing functions from $[0, \infty)$ to $[0, \infty)$ with
$\varphi(0) = \psi(0) = 0$ that are inverse to each other. Then for $a, b \geq 0$, we have

$$ab \leq \int_0^a \varphi(x)\,dx + \int_0^b \psi(y)\,dy,$$

with equality if and only if $b = \varphi(a)$. In Figure 8.20a, we illustrate the
inequality for $b > \varphi(a)$, and in Figure 8.20b for $b < \varphi(a)$ [Tolsted, 1964].

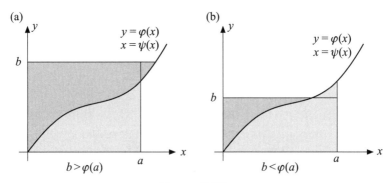

Figure 8.20.

Young's result is so general that many interesting inequalities may be derived from it. For example, when $\varphi(t) = \psi(t) = t$ (and replacing a and b
by their square roots) we obtain the AM-GM inequality:

$$\sqrt{ab} \leq \int_0^{\sqrt{a}} x\,dx + \int_0^{\sqrt{b}} y\,dy = \frac{a+b}{2}.$$

If we let $\varphi(x) = x^{p-1}$ and $\psi(y) = y^{1/(p-1)}$ for $p > 1$, then we have

$$ab \leq \int_0^a x^{p-1} dx + \int_0^b y^{1/(p-1)} dy = \frac{a^p}{p} + \frac{p-1}{p} b^{p/(p-1)}.$$

Letting $q = p/(p-1)$ yields a symmetric form of this inequality: for $p, q > 1$ with $(1/p) + (1/q) = 1$, we have

$$ab \leq \frac{a^p}{p} + \frac{b^q}{q} \tag{8.5}$$

with equality when $a^p = b^q$.

William Henry Young

Otto Ludwig Hölder

Hermann Minkowski

Figure 8.21.

Application 8.12. *Hölder's inequality* [Hölder, 1889]

Given two sets of real numbers $\{a_1, a_2, \ldots, a_n\}$, $\{b_1, b_2, \ldots, b_n\}$, let

$$A = \left(\sum_{i=1}^n |a_i|^p\right)^{1/p} \text{ and } B = \left(\sum_{i=1}^n |b_i|^q\right)^{1/q}.$$

Set $a = |a_i|/A, b = |b_i|/A$. Then by (8.5), for all i in $\{1, 2, \ldots, n\}$ and $p, q > 1$ such that $(1/p) + (1/q) = 1$, we have

$$\frac{|a_i|}{A} \cdot \frac{|b_i|}{B} \leq \frac{(|a_i|/A)^p}{p} + \frac{(|b_i|/B)^q}{q}.$$

Summing these inequalities on i from 1 to n yields

$$\frac{1}{AB} \sum_{i=1}^n |a_i b_i| \leq \frac{1}{pA^p} \sum_{i=1}^n |a_i|^p + \frac{1}{qB^q} \sum_{i=1}^n |b_i|^q = \frac{1}{p} + \frac{1}{q} = 1,$$

which proves *Hölder's inequality* [Otto Ludwig Hölder (1859–1937)]:

$$\sum_{i=1}^n |a_i b_i| \leq \left(\sum_{i=1}^n |a_i|^p\right)^{1/p} \left(\sum_{i=1}^n |b_i|^q\right)^{1/q}. \tag{8.6}$$

The special case $p = q = 2$ is the Cauchy-Schwarz inequality (7.9).

Example 8.9. *The classical ladder problem*

Virtually every modern calculus text has a version of the ladder problem:

A ladder is being carried down a hallway a feet wide. At the end of the hallway is a 90° turn into another hallway b feet wide. What is the length of the longest ladder that can be carried horizontally around the corner? (Ignore the width of the ladder.)

The situation is illustrated in Figure 8.22, where for convenience we have located an axis system at the corner. The exercise can be solved immediately without calculus by using Hölder's inequality.

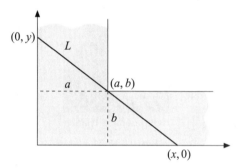

Figure 8.22.

The length of the ladder is $L = (x^2 + y^2)^{1/2}$, and by similar triangles $b/y = (x - a)/x$, i.e., $a/x + b/y = 1$. Using (8.6) with $n = 2$, $p = 3$, $q = 3/2$, $= x^{2/3}$, $a_2 = y^{2/3}$, $b_1 = (a/x)^{2/3}$, and $b_2 = (b/y)^{2/3}$ yields

$$\sum_{i=1}^{n} |a_i b_i| = x^{2/3} \left(\frac{a}{x}\right)^{2/3} + y^{2/3} \left(\frac{b}{y}\right)^{2/3} = a^{2/3} + b^{2/3},$$

$$\left(\sum_{i=1}^{n} |a_i|^p\right)^{1/p} = \left[(x^{2/3})^3 + (y^{2/3})^3\right]^{1/3} = (x^2 + y^2)^{1/3} = L^{2/3},$$

and

$$\left(\sum_{i=1}^{n} |b_i|^q\right)^{1/q} = \left[\left((a/x)^{2/3}\right)^{3/2} + \left((b/y)^{2/3}\right)^{3/2}\right]^{2/3} = 1.$$

Thus $a^{2/3} + b^{2/3} \le L^{2/3}$, or $L \ge \left(a^{2/3} + b^{2/3}\right)^{3/2}$, so the length of the longest ladder that can be carried around the corner is $L = \left(a^{2/3} + b^{2/3}\right)^{3/2}$.

Application 8.13. *Minkowski's inequality* [Minkowski, 1896]

For two sets of real numbers $\{a_1, a_2, \ldots, a_n\}$, $\{b_1, b_2, \ldots, b_n\}$, *Minkowski's inequality* [Hermann Minkowski (1864–1909)] states that for $p \geq 1$,

$$\left(\sum_{i=1}^{n} |a_i + b_i|^p\right)^{1/p} \leq \left(\sum_{i=1}^{n} |a_i|^p\right)^{1/p} + \left(\sum_{i=1}^{n} |b_i|^p\right)^{1/p}. \quad (8.7)$$

(The direction of the inequality reversed for $0 < p < 1$.) When $n = p = 1$, this is the triangle inequality, for $p = 1/2$ we have the special case illustrated in Section 1.2, and for $p = 2$ we have the triangle inequality for vectors illustrated in Application 7.4.

We will prove the case $p > 1$ using Hölder's inequality (for the general case, see [Hardy, Littlewood, and Pólya, 1959]). Let $q = p/(p-1)$. Then, using (8.6) twice, we have

$$\sum_{i=1}^{n} |a_i + b_i|^p \leq \sum_{i=1}^{n} |a_i||a_i + b_i|^{p-1} + \sum_{i=1}^{n} |b_i||a_i + b_i|^{p-1}$$

$$\leq \left(\sum_{i=1}^{n} |a_i|^p\right)^{1/p} \left(\sum_{i=1}^{n} |a_i + b_i|^{(p-1)q}\right)^{1/q}$$

$$+ \left(\sum_{i=1}^{n} |b_i|^p\right)^{1/p} \left(\sum_{i=1}^{n} |a_i + b_i|^{(p-1)q}\right)^{1/q}$$

$$\leq \left[\left(\sum_{i=1}^{n} |a_i|^p\right)^{1/p} + \left(\sum_{i=1}^{n} |b_i|^p\right)^{1/p}\right]$$

$$\times \left(\sum_{i=1}^{n} |a_i + b_i|^p\right)^{1/q}.$$

Regrouping terms yields the result.

8.8   Bounded monotone sequences

As a consequence of the monotone convergence theorem, every sequence $\{a_1, a_2, \ldots, a_n, \ldots\}$ of real numbers that is non-decreasing and bounded from above has a limit. Therefore to establish the convergence of $\{a_1, a_2, \ldots, a_n, \ldots\}$, one need only show that for all $n \geq 1$, $a_n \leq a_{n+1} \leq M$ for some M. Often the ideas of this chapter concerning the representation of the values of a_n and M as slopes of line segments or areas of figures may be a convenient tool for establishing the required inequalities.

Example 8.10. *The Euler-Mascheroni constant*

The well-known Euler-Mascheroni constant γ (whose irrationality is still unknown) may be defined by means of the expression [Dunham, 1999]

$$\gamma = \lim_{n\to\infty} \left(\sum_{k=1}^{n} (1/k) - \ln(n+1)\right). \quad (8.8)$$

In order to show that (8.8) makes sense, i.e., the limit *exists*, we consider the sequence $\{a_n\}$ with

$$a_n = \sum_{k=1}^{n} (1/k) - \ln(n+1)$$

for $n \geq 1$, and show that it is increasing and bounded from above. Now

$$a_{n+1} - a_n = \frac{1}{n+1} - \ln(n+2) + \ln(n+1) = \frac{1}{n+1} - \int_{n+1}^{n+2} \frac{1}{x}\, dx > 0,$$

as illustrated in Figure 8.23, where the region shaded gray represents the difference between the area $1/(n+1)$ of a rectangle and the area under the graph of $y = 1/x$ over the interval $[n+1, n+2]$.

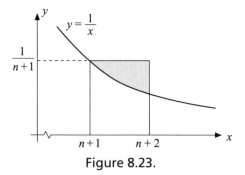

Figure 8.23.

Furthermore, $a_n < 1$ for all n, as shown in Figure 8.24, where a_n is represented by the n gray shaded regions above the graph of $y = 1/x$ over the interval $[1, n+1]$.

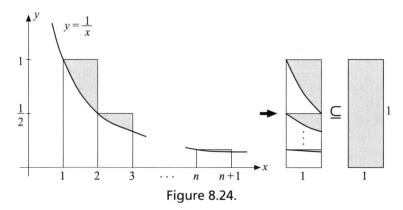

Figure 8.24.

Hence the limit in (8.8) exists, and evaluated to 20 decimal places, $\gamma \approx 0.57721566490153286060$.

8.9 Challenges

8.1 Let $f : [0, \infty) \to [0, \infty)$.

  (a) Prove that if f is concave, then f is subadditive.

  (b) Prove that if f is convex, then f is superadditive.

8.2 Use Jensen's inequality to give another proof of the power mean inequality (see Theorem 2.5).

8.3 Prove Lemma 8.1 and Theorem 8.2.

8.4 (a) Use the mediant property (Section 2.7) to establish (8.4).

  (b) Prove $\sum_{n=1}^{\infty} 1/\sqrt{n}$ diverges.

8.5 Show that the sequence $\{(1 + 1/n)^n\}_{n=1}^{\infty}$ converges. [Hint: consider tangents and secants to the graph of $y = \ln(1 + x)$.]

8.6 Are $f(x) = |\sin x|$ and $g(x) = |\cos x|$ subadditive on $(0, \infty)$?

8.7 Show, using a moving frame argument, that any function that satisfies a Lipschitz condition with constant M on an interval I must be uniformly continuous on I.

8.8 If $0 < a < b$ and $0 < t < 1$, show that $a^{1-t}b^t < (1 - t)a + tb$.

8.9 As a consequence of (1.1) and Challenge 1.4, we have for $0 < a < b$,

$$a < \frac{2ab}{a + b} < \sqrt{ab} < \frac{a + b}{2} < \sqrt{\frac{a^2 + b^2}{2}} < \frac{a^2 + b^2}{a + b} < b.$$

Illustrate these inequalities by superimposing the graphs of $y = x, x + y = a + b, xy = ab, x^2 + y^2 = (a + b)^2/2, x^2 + y^2 = a^2 + b^2$, and $xy = (a^2 + b^2)/2$.

8.10 Show that for any $a < b, (e^b - e^a)/(b - a) > e^{(a+b)/2}$.

8.11 Show that $e^x \geq 1 + x$ for all real x.

8.12 What inequality is illustrated in Figure 8.25?

8.13 Let $p, q > 1$ satisfy $1/p + 1/q = 1$, and let u, v be non-negative real numbers, at least one of them positive. Define f on $[0, 1]$ by $f(t) = ut + v(1 - t^q)^{1/q}$. Using its graph, show that f has an absolute maximum at the point $[u^p/(u^p + v^p)]^{1/q}$. Use this result to establish the $n = 2$ case of Hölder's inequality: $a_1b_1 + a_2b_2 \leq (a_1^p + a_2^p)^{1/p}(b_1^q + b_2^q)^{1/q}$.

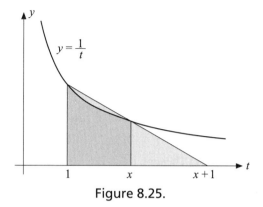

Figure 8.25.

8.14 Show that in any convex polyhedron with F faces, V vertices, and E edges, we have $E \le 3V - 6$ and $F \le 3V - 4$.

8.15 (a) Create a visual argument to show that the function $\sin x / x$ is non-increasing on $(0, \pi)$.

　　　(b) Show that the sequence $\{n \sin(1/n)\}_{n=1}^{\infty}$ converges.

8.16 Let $f : [0, \infty) \to [0, \infty)$ be a superadditive function. What are the geometric implications of this property for the graph of f?

CHAPTER 9

Additional topics

In this final chapter, we examine some methods for illustrating inequalities by combining two or more techniques from earlier chapters. We also give a brief introduction to the theory of majorization, which has proven to be remarkably effective in proving inequalities.

9.1 Combining inequalities

We have seen in previous chapters several examples where a final inequality is obtained after illustrating several simpler inequalities and then combining them. See, for example, Sections 4.1 and 4.3. We now revisit this methodology.

Example 9.1. *If a, b, c are the sides of a right triangle with hypotenuse c, then $c(a + b) \geq 2\sqrt{2}ab$.*

Consider the square in Figure 9.1. Comparing the area of the large square to the area of the four a by b rectangles yields $(a + b)^2 \geq 4ab$, or $a + b \geq 2\sqrt{ab}$ (our final proof of the AM-GM inequality, we promise). Comparing the area of the square with side c to the area of the four gray triangles yields

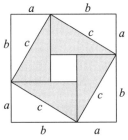

Figure 9.1.

$c^2 \geq 2ab$, or $c \geq \sqrt{2ab}$. Multiplying now yields $c(a + b) \geq 2\sqrt{2}ab$, as claimed.

Example 9.2. *A tetrahedral inequality*

Let $OABC$ be a tetrahedron in xyz-space with vertices $O = (0, 0, 0)$, $A = (a, 0, 0)$, $B = (0, b, 0)$, and $C = (0, 0, c)$, with a, b, c all positive. If V denotes the volume and L the sum of the lengths of the six edges of the tetrahedron, then

$$V \leq L^3 / (6(3 + 3\sqrt{2})^3).$$

Since $V = abc/6$, the inequality is equivalent to $3\sqrt[3]{abc} \leq L/(1 + \sqrt{2})$. Three applications of the arithmetic mean-root mean square inequality yield

$$a+b+c = \frac{a + b}{2} + \frac{b + c}{2} + \frac{c + a}{2} \leq \sqrt{\frac{a^2 + b^2}{2}} + \sqrt{\frac{b^2 + c^2}{2}} + \sqrt{\frac{c^2 + a^2}{2}},$$

whence

$$a + b + c \leq \frac{1}{\sqrt{2}} \left(\sqrt{a^2 + b^2} + \sqrt{b^2 + c^2} + \sqrt{c^2 + a^2} \right)$$

$$= \frac{1}{\sqrt{2}} (L - a - b - c),$$

so $(1 + \sqrt{2})(a + b + c) \leq L$. By virtue of the AM-GM inequality for a, b, c, we have

$$3\sqrt[3]{abc} \leq a + b + c \leq L/(1 + \sqrt{2}),$$

which establishes the desired result.

Example 9.3. *Bellman's inequality*

By combining Minkowski's inequality (see Application 8.13), the triangle inequality, and the Pythagorean relation for triangles, we can establish *Bellman's inequality*: if $a_1, a_2, \ldots, a_n$ and $b_1, b_2, \ldots, b_n$ are nonnegative real numbers such that $a_1^2 - a_2^2 - a_3^2 - \cdots - a_n^2 > 0$ and $b_1^2 - b_2^2 - b_3^2 - \cdots - b_n^2 > 0$, then

$$\sqrt{a_1^2 - a_2^2 - \cdots - a_n^2} + \sqrt{b_1^2 - b_2^2 - \cdots - b_n^2}$$
$$\leq \sqrt{(a_1 + b_1)^2 - (a_2 + b_2)^2 - \cdots - (a_n + b_n)^2}.$$

If $A < a_1$ and $B < b_1$, then as seen in Figure 9.2, we have

$$\sqrt{a_1^2 - A^2} + \sqrt{b_1^2 - B^2} = \sqrt{c^2 - (A + B)^2} \leq \sqrt{(a_1 + b_1)^2 - (A + B)^2}.$$

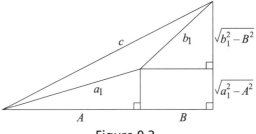

Figure 9.2.

Now set $A^2 = a_2^2 + a_3^2 + \cdots + a_n^2$ and $B^2 = b_2^2 + b_3^2 + \cdots + b_n^2$. This establishes the first inequality below and the second follows from Minkowski's inequality (8.7) with $p = 2$:

$$\sqrt{a_1^2 - a_2^2 - \cdots - a_n^2} + \sqrt{b_1^2 - b_2^2 - \cdots - b_n^2}$$
$$\leq \sqrt{(a_1 + b_1)^2 - \left(\sqrt{a_2^2 + \cdots + a_n^2} + \sqrt{b_2^2 + \cdots + b_n^2}\right)^2}$$
$$\leq \sqrt{(a_1 + b_1)^2 - \left(\sqrt{(a_2 + b_2)^2 + \cdots + (a_n + b_n)^2}\right)^2}$$
$$= \sqrt{(a_1 + b_1)^2 - (a_2 + b_2)^2 - \cdots - (a_n + b_n)^2}.$$

Application 9.1. *Oppenheim's inequality for two triangles*

An interesting result of A. Oppenheim is the following: let $A_i B_i C_i$ ($i = 1, 2$) be two triangles with sides a_i, b_i, c_i and areas K_i. Let $A_3 B_3 C_3$ be a triangle with sides $a_3 = \sqrt{a_1^2 + a_2^2}$, $b_3 = \sqrt{b_1^2 + b_2^2}$, and $c_3 = \sqrt{c_1^2 + c_2^2}$, and area K_3. Then we have the inequality $K_1 + K_2 \leq K_3$ (with equality if and only if the triangles are similar). As remarked by Mitrinović [Mitrinović et al., 1989], this geometric inequality is a consequence of Bellman's inequality with (a_1, a_2, a_3, a_4) replaced by $(a_1^2 + b_1^2 + c_1^2, \sqrt{2}a_1^2, \sqrt{2}b_1^2, \sqrt{2}c_1^2)$, (b_1, b_2, b_3, b_4) replaced by $(a_2^2 + b_2^2 + c_2^2, \sqrt{2}a_2^2, \sqrt{2}b_2^2, \sqrt{2}c_2^2)$, and Heron's formula (see Application 4.1) where $16K_i^2 = (a_i^2 + b_i^2 + c_i^2)^2 - 2(a_i^4 + b_i^4 + c_i^4)$. Note the similarity to the connection between the Neuberg-Pedoe inequality (Application 7.6) and Aczél's inequality (Section 7.4).

9.2 Majorization

Any book on inequalities would be incomplete without a discussion of the theory of majorization. Although the theory does not make extensive use of visual techniques it does have important applications in geometry. As

A. Marshall and I. Olkin note in the Preface to their seminal work *Inequalities: Theory of Majorization and its Applications* [Marshall and Olkin, 1979]:

> Although they play a fundamental role in nearly all branches of mathematics, inequalities are usually obtained by ad hoc methods rather than as consequences of some underlying "theory of inequalities." For certain kinds of inequalities, the notion of majorization leads to such a theory that is sometimes extremely useful and powerful for deriving inequalities.

The theory of majorization requires advanced algebraic and functional machinery that is beyond the scope of this book, but we will present some basic definitions, examples and applications without proofs.

Majorization is a pre-order $\prec$ on the set of n-dimensional vectors. Let $\mathbf{x} = (x_1, x_2, \ldots, x_n)$, $\mathbf{y} = (y_1, y_2, \ldots, y_n)$, let $x_{[1]} \geq x_{[2]} \geq \cdots \geq x_{[n]}$ denote the components of $\mathbf{x}$ in nonincreasing order, and let $y_{[1]} \geq y_{[2]} \geq \cdots \geq y_{[n]}$ denote the components of $\mathbf{y}$ in nonincreasing order. Then $\mathbf{x}$ is *majorized* by $\mathbf{y}$, written $\mathbf{x} \prec \mathbf{y}$, if the sum of the k largest components of $\mathbf{x}$ is less than or equal to the sum of the k largest components of $\mathbf{y}$, $k = 1, 2, \ldots, n$ with equality when $k = n$. Equivalently, $\mathbf{x} \prec \mathbf{y}$ if

$$\sum_{i=1}^{k} x_{[i]} \leq \sum_{i=1}^{k} y_{[i]} \quad \text{for} \quad k$$
$$= 1, 2, \ldots, n-1; \text{ and } \sum_{i=1}^{n} x_{[i]} = \sum_{i=1}^{n} y_{[i]}.$$

For example, when $n = 4$,

$$\left(\frac{1}{4}, \frac{1}{4}, \frac{1}{4}, \frac{1}{4}\right) \prec \left(\frac{1}{3}, \frac{1}{3}, \frac{1}{3}, 0\right) \prec \left(\frac{1}{2}, \frac{1}{2}, 0, 0\right) \prec (1, 0, 0, 0).$$

So, in a sense, $\mathbf{x} \prec \mathbf{y}$ if the components of $\mathbf{x}$ are less spread out than the components of $\mathbf{y}$. When discussing majorization, we usually write the components of a vector in nonincreasing order. Majorization fails to be a partial order since $\mathbf{x} \prec \mathbf{y}$ and $\mathbf{y} \prec \mathbf{x}$ do not imply $\mathbf{x} = \mathbf{y}$, but only that the elements of $\mathbf{y}$ are a permutation of the elements of $\mathbf{x}$.

Many inequalities relating three parameters (angles, sides, altitudes, ...) of a triangle have succinct representations in terms of majorization. For instance, for the angles A, B, C of a triangle, we have

$$\left(\frac{\pi}{3}, \frac{\pi}{3}, \frac{\pi}{3}\right) \prec (A, B, C) \prec (\pi, 0, 0) \text{ for all triangles,} \tag{9.1}$$

$$\left(\frac{\pi}{3}, \frac{\pi}{3}, \frac{\pi}{3}\right) \prec (A, B, C) \prec \left(\frac{\pi}{2}, \frac{\pi}{2}, 0\right) \text{ for acute triangles, and} \tag{9.2}$$

$$\left(\frac{\pi}{2}, \frac{\pi}{4}, \frac{\pi}{4}\right) \prec (A, B, C) \prec (\pi, 0, 0) \text{ for obtuse triangles.} \tag{9.3}$$

In Figure 9.3a, we illustrate $(\pi/3, \pi/3, \pi/3) \prec (A, B, C)$ by showing that for an arbitrary triangle with angles $A \geq B \geq C$, we have $A \geq \pi/3$ and $A + B \geq 2\pi/3$, and in Figure 9.3b, we illustrate $(\pi/2, \pi/4, \pi/4) \prec (A, B, C)$ for obtuse triangles with angles $A \geq B \geq C$ by showing that $A \geq \pi/2$ and $A + B \geq 3\pi/4$. The triangles whose sides are dashed lines are equilateral triangles in (a) and isosceles right triangles in (b).

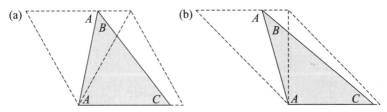

Figure 9.3.

Functions that preserve majorization order are called Schur-convex. If S is a set of n-dimensional vectors, then $F : S \rightarrow (-\infty, \infty)$ is *Schur-convex* if $\mathbf{x} \prec \mathbf{y}$ implies $F(\mathbf{x}) \leq F(\mathbf{y})$. Similarly, $F : S \rightarrow (-\infty, \infty)$ is *Schur-concave* if $\mathbf{x} \prec \mathbf{y}$ implies $F(\mathbf{x}) \geq F(\mathbf{y})$. A classical result of I. Schur [Hardy, Littlewood and Pólya, 1959] states that if g is convex (concave) on an interval I, then $F(\mathbf{x}) = g(x_1) + g(x_2) + \cdots + g(x_n)$ is Schur-convex (Schur-concave) on $S = I^n$.

For example, since the sine function is concave on $(0, \pi)$, we immediately obtain from (9.1), (9.2), and (9.3) the following inequalities:

$$0 < \sin A + \sin B + \sin C \leq 3\sqrt{3}/2 \text{ for all triangles,}$$

$$2 < \sin A + \sin B + \sin C \leq 3\sqrt{3}/2 \text{ for acute triangles, and}$$

$$0 < \sin A + \sin B + \sin C \leq 1 + \sqrt{2} \text{ for obtuse triangles.}$$

Using the concavity of functions such as $\sqrt{\sin x}$, $\ln(\sin x)$, $\sin(kx)$ on $(0, \pi/k)$, or the convexity of $\sin^2(x/2)$ on $(0, \pi/2)$, we can derive a host of similar inequalities for the angles of a triangle.

Application 9.2. *Flanders' inequality*

Flanders' inequality relates the product of the sines of the angles A, B, C of a triangle to the product of the magnitudes of the angles (in radians):

$$\sin A \sin B \sin C \leq \left(\frac{3\sqrt{3}}{2\pi}\right)^3 ABC.$$

Let $g(x) = \ln(\sin x/x)$. Using elementary calculus, we can show that g is concave on $(0, \pi)$, and hence $(\pi/3, \pi/3, \pi/3) \prec (A, B, C)$ implies that

$$\ln \frac{\sin A \sin B \sin C}{ABC} \le \ln \left(\frac{\sqrt{3}/2}{\pi/3} \right)^3,$$

from which Flanders' inequality follows.

If a, b, c are the sides and s the semiperimeter of a triangle, then

$$\left(\frac{2}{3}s, \frac{2}{3}s, \frac{2}{3}s \right) \prec (a, b, c) \prec (s, s, 0), \tag{9.4}$$

where the first majorization holds for all numbers a, b, c. The second requires that a, b, c be the sides of a triangle (i.e., it is equivalent to the triangle inequality).

For example, the convexity of $g(x) = x^2$ immediately establishes $3(2s/3)^2 \le a^2 + b^2 + c^2 \le 2s^2$, or

$$2(a^2 + b^2 + c^2) \le (a + b + c)^2 \le 3(a^2 + b^2 + c^2). \tag{9.5}$$

The first inequality in (9.5) is illustrated in Figure 9.4 (in the form $a^2 + b^2 + c^2 \le (1/2)(a + b + c)^2$), assuming $a \ge b \ge c$ and the triangle inequality $a < b + c$. The second inequality in (9.5) is equivalent to the one in Lemma 2.1 and illustrated in Figure 2.8.

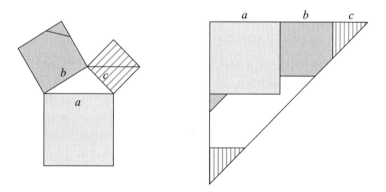

Figure 9.4.

Application 9.3.   *The isoperimetric inequality for triangles revisited*

In Challenge 4.3 we encountered the isoperimetric inequality for triangles: if K denotes the area and L the perimeter of a triangle, then $L^2 \ge 12\sqrt{3}K$, or $s^2 \ge 3\sqrt{3}K$ where s denotes the semiperimeter. Let $g(x) =$

$\ln(s - x)$. Since g is concave on $(0, s)$, it follows from the first majorization in (9.4) that $\ln (s/3)^3 \geq \ln[(s - a)(s - b)(s - c)]$, so

$$\frac{s^4}{27} \geq s(s - a)(s - b)(s - c) = K^2,$$

from which the desired result follows.

Many other applications of majorization to inequalities involving the elements in a triangle can be found in [Marshall and Olkin, 1979].

9.3 Challenges

9.1 Draw a picture similar to Figure 9.3 to illustrate $(A, B, C) \prec (\pi/2, \pi/2, 0)$ for acute triangles.

9.2 The substitution $a = x + y, b = y + z, c = z + x$ for the sides a, b, c of a triangle is called the Ravi substitution (see Section 1.6).

(a) Show that $(a, b, c) \prec (2y, 2x, 2z) \prec (2s, 0, 0)$.

(b) Use (a) to establish Padoa's inequality (1.2).

(c) Show that $\frac{1}{a} + \frac{1}{b} + \frac{1}{c} \leq \frac{1}{b+c-a} + \frac{1}{a+c-b} + \frac{1}{a+b-c}$.

9.3 Find visual proof(s) of the inequality

$$a^k (a - b) + b^k (b - c) + c^k (c - a) \geq 0$$

where $k > 0$ and a, b, c are the sides of a triangle. When do we have equality?

9.4 If a, b, c are the sides of a triangle with semiperimeter s, show that

$$\sqrt{(s - b)(s - c)} + \sqrt{bc} \leq s.$$

9.5 Create a visual proof to show that if $a \geq b \geq c > 0$, then $a^2 + bc \geq a(b + c)$.

9.6 Let a, b, c be the sides of a triangle. Find visual proofs for:

(a) $(a + b + c)^2 \leq 4(ab + bc + ca)$,

(b) $3(ab + bc + ca) \leq (a + b + c)^2$,

(c) $(a + b + c)(a^2 + b^2 + c^2) \leq 3(a^3 + b^3 + c^3)$.

9.7 Can Bellman's inequality be an equality?

9.8 Let a, b, c denote the sides of triangle ABC with $a \geq b \geq c$, and prove the following inequalities:

(a) $\frac{C}{c} \leq \frac{B}{b} \leq \frac{A}{a}$,

(b) $(a - b) \left(\frac{A}{a} - \frac{B}{b} \right) + (b - c) \left(\frac{B}{b} - \frac{C}{c} \right) + (c - a) \left(\frac{C}{c} - \frac{A}{a} \right) \geq 0$,

(c) $2(A + B + C) \geq (b + c)\frac{A}{a} + (c + a)\frac{B}{b} + +(a + b)\frac{C}{c}$,

(d) $\frac{A}{a} + \frac{B}{b} + \frac{C}{c} \leq \frac{3\pi}{2s}$.

9.9 Show that in any acute triangle, $\frac{a}{A} + \frac{b}{B} + \frac{c}{C} > \frac{12R}{\pi}$.

9.10 Create a visual proof of the inequality for real numbers $x, y, z > 0$:

$(x + y + z) \left(\frac{1}{x} + \frac{1}{y} + \frac{1}{z} \right) \geq 9.$

Solutions to the Challenges

Many of the Challenges in this book have multiple solutions. So here we give but one simple solution to each Challenge, and encourage readers to search for others that may be simpler.

Chapter 1

1.1 (a) Extend CD to meet AB at a point X between A and B. Then by the triangle inequality, $|AD| \leq |AX| + |DX|$ and $|DC| + |DX| \leq |XB| + |BC|$, from which $|AD| + |CD| \leq |AB| + |BC|$ follows.

(b) Label as Y the point where EH and FG intersect. Then by the triangle inequality, $|EF| \leq |FY| + |EY|$ and $|GH| \leq |GY| + |HY|$, from which the desired result follows.

1.2 (a) The harmonic mean-geometric mean inequality.

(b) The sum of a positive number and its reciprocal is at least 2.

(a)   (b)

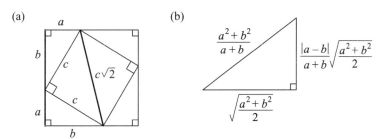

Figure S.1.

1.3 See Figure S.1a.

1.4 (a) See Figure S.1b.

(b) $\frac{a+b}{2} - \frac{2ab}{a+b} = \frac{(a-b)^2}{2(a+b)} = \frac{a^2+b^2}{a+b} - \frac{a+b}{2}$.

(c) The Heronian mean $\frac{a+\sqrt{ab}+b}{3}$ equals $\frac{2}{3}\frac{a+b}{2} + \frac{1}{3}\sqrt{ab}$.

1.5 Construct the quadrilateral $OPQR$ shown in Figure S.2. By the triangle
inequality, $|PQ|+|QR| \geq |PR|$. But from the law of cosines, $|PQ|^2 = a^2 + b^2 - ab$, $|QR|^2 = b^2 + c^2 - bc$, and $|PR|^2 = a^2 + c^2 + ac$.

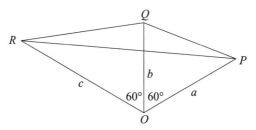

Figure S.2.

1.6 See Figure S.3a.

1.7 See Figure S.3b. The tangent line to $y = e^{x/e}$ at (e, e) is $y = x$, so
$e^{x/e} \geq x$; or $e^{1/e} \geq x^{1/x}$.

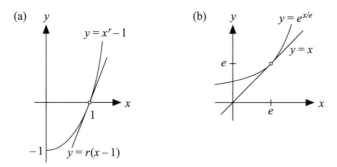

Figure S.3.

1.8 Let x, y, z denote respectively the length, width, and height of the box.
Then the amount of material is $S = xy + 2xz + 3yz$, and the volume
is $V = xyz$. But by the AM-GM inequality, we have

$$6V^2 = xy \cdot 2xz \cdot 3yz \leq \left(\frac{xy + 2xz + 3yz}{3} \right)^3 = \frac{1}{27} S^3,$$

with equality if and only if $xy = 2xz = 3yz$, or $x = 3z$ and $y = 2z$.
Hence the best design has the length and width 3 and 2 times the height,
respectively.

1.9 (a) Using the hint, the inequality is equivalent to $\frac{1}{x}+\frac{1}{y}+\frac{1}{z}\geq\frac{9}{x+y+z}$, which is true by the harmonic mean-arithmetic mean inequality applied to x, y, and z.

(b) Applying the Ravi substitution yields the double inequality $\sqrt{x+y+z}\leq\sqrt{x}+\sqrt{y}+\sqrt{z}\leq\sqrt{3(x+y+z)}$. The first inequality is equivalent to $x+y+z\leq x+y+z+2(\sqrt{xy}+\sqrt{yz}+\sqrt{zx})$, which is clearly true. The second inequality is equivalent to $\sqrt{xy}+\sqrt{yz}+\sqrt{zx}\leq x+y+z=\frac{x+y}{2}+\frac{y+z}{2}+\frac{x+z}{2}$, which is true by the AM-GM inequality.

1.10 Assume $a\leq b$. Locate D on side AC so $|CD|=a$ and $|AD|=b-a$, and extend CB to E so $|BE|=b-a$ and $|CE|=b$. Let $x=|BD|$, then $|AE|=bx/a$. See Figure S.4.

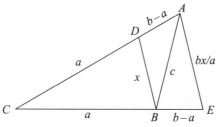

Figure S.4.

From the law of cosines, $x^2=a^2+a^2-2a^2\cos C$, so $x=a\sqrt{2-2\cos C}$. In $\triangle ABD$, $c\leq b-a+x=a+b-(2a-x)$, hence $c\leq a+b-(2-\sqrt{2-2\cos C})a$. In $\triangle ABE$, $bx/a\leq c+(b-a)$, or $c\geq a-b+bx/a=a+b-(2b-bx/a)=a+b-(2-\sqrt{2-2\cos C})b$.

1.11 Since $a_1=2<2\sqrt{2}=a_2$, we will assume $n\geq 2$. If one inscribes a regular 2^n-gon in a unit circle, the length of each side is $2\sin(\pi/2^n)$, hence its perimeter is $P_n=2^{n+1}\sin(\pi/2^n)=2a_n$. Since the regular 2^n-gon can be inscribed inside the regular 2^{n+1}-gon (sharing half its vertices), $P_n\leq P_{n+1}$, thus $a_n\leq a_{n+1}$. An upper bound for the perimeters is 2π, hence $a_n\leq\pi$ for all n.

Chapter 2

2.1 Let $x=$ length and $2y+2z=$ girth. If V denotes the volume, then $V=xyz=\frac{x\cdot 2y\cdot 2z}{4}\leq\frac{1}{4}\left(\frac{x+2y+2z}{3}\right)^3=\frac{1}{4}\left(\frac{165}{3}\right)^3$ with equality if and only if $x=2y=2z$. Hence $x=55$ inches and $y=z=27.5$ inches.

2.2 The AM-GM inequality for 2^k numbers $a_1, a_2, \ldots, a_{2^k}$ is $(a_1 a_2 \ldots a_{2^k})^{1/2^k} \leq (a_1 + a_2 + \cdots + a_{2^k})/2^k$. When $k = 1$ this is the AM-GM inequality for two numbers. Assume it is true for $k = n$. Then

$$(a_1 a_2 \ldots a_{2^n} a_{2^n+1} \ldots a_{2^{n+1}})^{1/2^{n+1}}$$

$$= \sqrt{(a_1 a_2 \cdots a_{2^n})^{1/2^n} (a_{2^n+1} \cdots a_{2^{n+1}})^{1/2^n}}$$

$$\leq \frac{1}{2} \left[(a_1 a_2 \cdots a_{2^n})^{1/2^n} + (a_{2^n+1} \cdots a_{2^{n+1}})^{1/2^n} \right]$$

$$\leq \frac{1}{2^{n+1}} (a_1 + a_2 + \cdots + a_{2^n} + a_{2^n+1} + \cdots + a_{2^{n+1}}).$$

Thus the AM-GM inequality holds for 2^k numbers, $k = 1, 2, \cdots$.

2.3 (a) Schur's inequality with $r = 1$ states

$$x(x - y)(x - z) + y(y - x)(y - z) + z(z - x)(z - y) \geq 0$$

so

$$x^3 + y^3 + z^3 + 3xyz \geq xy(x + y) + yz(y + z) + xz(x + z).$$

(b) From (a), we have

$$xyz \geq xy(x + y) + yz(y + z) + xz(x + z)$$
$$- x^3 - y^3 - z^3 - 2xyz$$
$$= (x + y - z)(y + z - x)(z + x - y).$$

2.4 (a) Each diagonal of Q splits Q into two triangles. If a triangle has sides a and b and included angle θ, then its area is $ab \sin \theta / 2 \leq ab/2$. Thus $K \leq (ab + cd)/2$ and $k \leq (ad + bc)/2$.

(b) From (a), $2K \leq (ab + cd + ad + bc)/2 = (a + c)(b + d)/2$.

(c) By the AM-GM inequality, $(a+c)(b+d) \leq ((a + b + c + d)/2)^2$, so from (b), $K \leq (a + b + c + d)^2/16$.

(d) By the AM-RMS inequality, $(a + b + c + d)/4 \leq \sqrt{(a^2 + b^2 + c^2 + d^2)/4}$, hence $K \leq (a^2 + b^2 + c^2 + d^2)/4$.

2.5 (a) The mediant of a/c and b/c is $\frac{a+b}{2c} = \frac{(a/c)+(b/c)}{2}$.

(b) The mediant of c/a and c/b is $\frac{2c}{a+b} = \frac{2c^2/ab}{(c/a)+(c/b)}$.

2.6 Without loss of generality, assume $a_1/b_1 \leq a_2/b_2 \leq \cdots \leq a_n/b_n$, and interpret the fractions as slopes of line segments, as illustrated in Figure S.5.

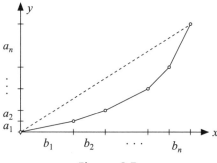

Figure S.5.

Hence

$$\frac{a_1}{b_1} \leq \frac{a_1 + a_2 + \ldots + a_n}{b_1 + b_2 + \ldots + b_n} \leq \frac{a_n}{b_n}.$$

2.7 In figure S.6 we have $1 - \sqrt{xy} \geq \sqrt{(1-x)(1-y)}$. Note that the AM-GM inequality insures that the length of the vertical side is real.

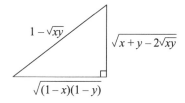

Figure S.6.

An algebraic solution can be obtained by applying the AM-GM inequality to both $\sqrt{xy}$ and $\sqrt{(1-x)(1-y)}$.

2.8 (a) The inequality is trivial if $a_1 = b_1 = 0$ or $a_2 = b_2 = 0$, so assume $a_1 + b_1 > 0$ and $a_2 + b_2 > 0$. Then divide both sides by $\sqrt{(a_1 + b_1)(a_2 + b_2)}$ to obtain

$$\sqrt{\frac{a_1}{a_1 + b_1} \cdot \frac{a_2}{a_2 + b_2}} + \sqrt{\frac{b_1}{a_1 + b_1} \cdot \frac{b_2}{a_2 + b_2}} \leq 1,$$

which is true by Challenge 2.7.

(b) As in (a), the inequality is trivial if $a_n = b_n = 0$ for any n, so assume $a_n + b_n > 0$ for all n, divide both sides by $\sqrt[n]{(a_1 + b_1)(a_2 + b_2) \cdots (a_n + b_n)}$, and then apply the AM-GM inequality:

$$\left(\frac{a_1}{a_1 + b_1} \cdot \frac{a_2}{a_2 + b_2} \cdots \frac{a_n}{a_n + b_n} \right)^{1/n}$$
$$\leq \frac{1}{n} \left(\frac{a_1}{a_1 + b_1} + \frac{a_2}{a_2 + b_2} + \cdots + \frac{a_n}{a_n + b_n} \right);$$

$$\left(\frac{b_1}{a_1 + b_1} \cdot \frac{b_2}{a_2 + b_2} \cdots \frac{b_n}{a_n + b_n} \right)^{1/n}$$
$$\leq \frac{1}{n} \left(\frac{b_1}{a_1 + b_1} + \frac{b_2}{a_2 + b_2} + \cdots + \frac{b_n}{a_n + b_n} \right).$$

When the inequalities are added, the right-hand side simplifies to 1.

2.9 Apply the HM-AM inequality to $1/(a + b)$, $1/(b + c)$, and $1/(c + a)$.

2.10 Apply the AM-GM inequality to $\{1, 2, 3, \ldots, n\}$.

2.11 If the dimensions of the box are x, y, and z, then the volume $V = xyz$ and the surface area $S = 2(xy + yz + xz)$. Hence

$$V = \sqrt{xy \cdot yz \cdot zx} \leq \left(\frac{xy + yz + zx}{3} \right)^{3/2} = (S/6)^{3/2},$$

with equality if and only if $x = y = z$. Hence if V is a constant, S is minimal for a cube.

2.12 We compare the squares of $\sqrt{10} + \sqrt{18}$ and $\sqrt{12} + \sqrt{15}$ in Figure S.7. The sums of the shaded rectangles in each square are the same, and since $10 + 18 > 12 + 15$, we conclude $\sqrt{10} + \sqrt{18} > \sqrt{12} + \sqrt{15}$.

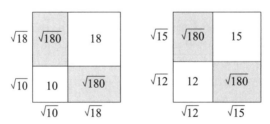

Figure S.7.

2.13 See Figure S.8 [Jiang, 2007].

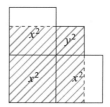

Figure S.8.

2.14 See Challenge 2.3(b).

2.15 Applying (2.1b) to the sets $\{x - 1, x, x + 1\}$ and $\{1/(x - 1), 1/x, 1/(x+1)\}$ yields $(3x)\left(\frac{1}{x-1} + \frac{1}{x} + \frac{1}{x+1}\right) > 3 \cdot 3$, from which Mengoli's inequality follows.

2.16 See Figure S.9.

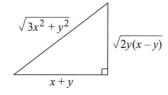

Figure S.9.

For a proof using Chapter 1 techniques, see Figure S.10.

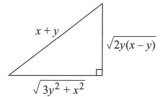

Figure S.10.

2.17 No. If the length of a side of the triangle is s, then its area is $s^2\sqrt{3}/4$, while the area of the square is $4s^2/9$. Since $\sqrt{3}/4 < 4/9$, the square is too large.

2.18 See Figure S.11.

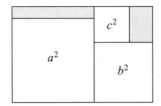

Figure S.11.

Note that a, b, c do not need to be the sides of a triangle.

2.19 Since $a^2 + b^2 = c^2$, we need only prove $ac - a^2 \leq c^2/4$. But by the AM-GM inequality, $ac = 2 \cdot a \cdot (c/2) \leq a^2 + c^2/4$, which establishes the result.

2.20 The AM-GM inequality.

2.21 (a) Compare the slope of the segment joining $(-x, 0)$ and (a, a) to the slope of the segment joining $(-x, 0)$ and (b, b).

 (b) We need only show that $a < b + c$ implies $\frac{a}{a+1} < \frac{b}{b+1} + \frac{c}{c+1}$. Since $a < b + c < b + c + bc$, part (a) with $x = 1$ yields

$$\frac{a}{a+1} < \frac{b+c+bc}{b+c+bc+1} < \frac{b+bc+c+bc}{(b+1)(c+1)} = \frac{b}{b+1} + \frac{c}{c+1}.$$

Chapter 3

3.1 Since $\triangle ABC$ is acute, each angle is less than $\pi/2$, so $A < \pi/2 < B + C$, etc.

3.2 We need $h_a > 2r$ and $h_a - r \leq r/\sin(A/2)$. See Figure S.12.

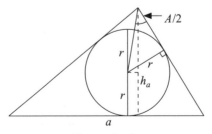

Figure S.12.

3.3 See Figure 3.1 to conclude that $b \geq h_a > 0$ and $c \geq h_a > 0$ are necessary and sufficient conditions for the existence of the triangle. For a, h_a and h_b we need $a \geq h_b$ (see Figure 3.3).

3.4 Draw a figure to see that a, $2m_b/3$, and $2m_c/3$ must form a triangle, and write the necessary (and sufficient) inequalities.

3.5 Draw the triangles with sides a, R, R and b, R, R; then construct the triangle with sides a, b, c. This construction is possible if $a < 2R$ and $b < 2R$.

3.6 First, we need $a < 2R$ and $b < 2R$. But a, $b/2$, and m_b must also form a triangle, so we also need $m_b < b/2 + a$, $b/2 < m_b + a$, and $a < b/2 + m_b$.

3.7 Clearly $a < b + c$ is required. In addition, from the law of sines, we also need $b + c = \frac{a}{\sin(A/2)} \sin(B + A/2) \leq \frac{a}{\sin(A/2)}$. See Figure S.13.

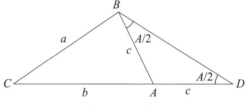

Figure S.13.

3.8 Since $\sin A = a/2R$, $\sin B = b/2R$, and $\sin C = c/2R$, the triangle inequalities of the form $a < b + c$ yield the corresponding inequalities for $\sin A$, $\sin B$, and $\sin C$. Note that $\cos A$, $\cos B$, and $\cos C$ do not necessarily determine a triangle since $\cos A \leq 0$ if $A \geq \pi/2$.

3.9 On a segment of length a, draw the arc of the circle as shown in Figure S.14, so the central angle subtending the given segment has measure $2A$. Hence any angle with vertex on the arc has measure A. So choose a point on the arc whose vertical height above the base line is h_a. This requires that $h_a \leq a \cot(A/2)/2$.

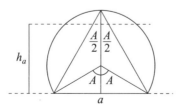

Figure S.14.

3.10 Since the area K of the triangle satisfies $2K = ah_a = bh_b = ch_c$, we have $a^2h_a = 2Ka$, $b^2h_b = 2Kb$, and $c^2h_c = 2Kc$. Thus the desired triangle not only exists, but will be similar to the original one.

3.11
$$w_a = w_b \Rightarrow \frac{2\sqrt{bc}}{b+c}\sqrt{s(s-a)} = \frac{2\sqrt{ac}}{a+c}\sqrt{s(s-b)}$$

$$\Rightarrow \frac{b(b+c-a)}{(b+c)^2} = \frac{a(a+c-b)}{(a+c)^2}$$

$$\Rightarrow (b-a)\big[(a+b+c)(c^2+ab) + 2abc\big] = 0$$

$$\Rightarrow a = b.$$

Chapter 4

4.1 (a) By Euler's triangle inequality and Lemmas 4.1 and 4.2, $abcs = 4RKs \geq 8rKs = 8K^2$.

   (b) Follows from (a) and Lemma 4.3.

   (c) Follows from Lemma 4.3 and Euler's triangle inequality.

   (d) By Lemmas 4.1 and 4.3, $abc = 4KR = 4Rrs$, so by the AM-GM inequality, $(2s)^3 = (a+b+c)^3 \geq 27abc = 27 \cdot 4Rrs$, hence $2s^2 \geq 27Rr$.

   (e By Heron's formula, Lemma 4.3, and the AM-GM inequality, we have

$$r^2s^2 = K^2 = s(s-a)(s-b)(s-c)$$

$$\leq s\left(\frac{s-a+s-b+s-c}{3}\right)^3 = s^4/27.$$

4.2 See Figure S.15.

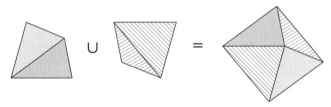

Figure S.15.   ·

4.3 (a) From Corollary 2.1, $R^2 \geq 4K/3\sqrt{3}$ (with equality if and only if the triangle is equilateral); so Lemma 4.1 implies that $(abc)^2 = 16K^2R^2 \geq 64K^3/3\sqrt{3}$, hence $K^3 \leq 3\sqrt{3}(abc)^2/64$.

   (b) From (a) and the AM-GM inequality, we have $(a + b + c)^2 \geq (3\sqrt[3]{abc})^2 = 9(abc)^{2/3} \geq 9 \cdot 4K/\sqrt{3} = 12\sqrt{3}K$, with equality if and only if $a = b = c$.

4.4 See Challenge 2.4(c).

4.5 In Figure 4.8 we showed that $K \leq pq/2$ with equality if and only if the diagonals of Q are perpendicular. But from the AM-GM inequality, $pq \leq (p^2 + q^2)/2$ with equality if and only if $p = q$.

4.6 Using the hint, the n segments form n isosceles triangles, which can be rearranged to form a second cyclic n-gon with the same area.

4.7 (a) From the proof of Lemma 4, 4, we see that K is maximized when the diagonals are perpendicular and equal in length to the diameter of the circumcircle. In such a case, the quadrilateral is a square.

   (b) In a bicentric quadrilateral, $a + c = b + d$, so by Brahmagupta's formula $K = \sqrt{abcd}$.

4.8 Squaring both sides of the inequality, bringing all terms to the right side, and factoring yields $0 < (a+b+c)(-a+b+c)(a-b+c)(a+b-c)$. If a, b, c form a triangle, then all the factors in the right side of the above inequality are positive. If the inequality is true and $a + b + c > 0$, then the other three factors are positive, since the possibility of two negative factors yields the contradiction that at least one of a, b, c must be negative.

4.9 See Figure S.16.

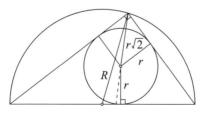

Figure S.16.

4.10 Since $K = rs = ah_a/2 = bh_b/2 = ch_c/2$ and $K = abc/4R$, Euler's inequality $R = abc/4rs \geq 2r$ yields

$$\frac{2rs}{h_a} \cdot \frac{2rs}{h_b} \cdot \frac{2rs}{h_c} \Big/ 4r \left(\frac{rs}{h_a} + \frac{rs}{h_b} + \frac{rs}{h_c} \right) \geq 2r,$$

which simplifies to the desired inequality.

4.11 In each part, apply the AM-GM inequality and the fact that $K = \sqrt{abcd}$ (see Challenge 4.7(b)).

4.12 Draw the hexagon connecting the centers of the circles to help in the computation of the perimeters P. See Figure S.17.

(a)   (b)   (c)   (d)   (e)

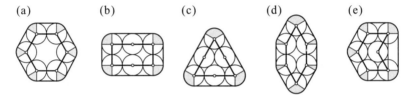

Figure S.17.

The first four cases have $P = 12 + 2\pi$, which is greater than $P = 8 + 2\sqrt{3} + 2\pi$ in case (e).

Chapter 5

5.1 From the proof of Theorem 3.3, we have $c = x + y$, with $x = ac/(a + b)$ and $y = bc/(b + c)$. By the triangle inequality, $|AA'| \leq 2b$. Since $\triangle BDC$ is similar to $\triangle BAA'$, $w_c/x = |AA'|/c \leq 2b/c$, so $w_c \leq \frac{2b}{c} \cdot \frac{ac}{a+b} = \frac{2ab}{a+b}$, as claimed.

5.2 We claim the path illustrated in Figure S.18a has minimal length. To prove this, let $AXYB$ be another path with $|XY| = s$. Then we have (see Figure S.18b)

$$|AX| + |XY| + |YB| = |A'X| + |XY| + |XB'|$$
$$\geq |A'B'| + |PQ| = |AP| + |PQ| + |QB|.$$

(a)   (b)

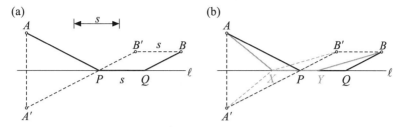

Figure S.18.

5.3 Reflections of P in each side of the angle produce the points P' and P'', and the line $P'P''$ cuts the sides in the desired points Q and R. For any other points Q' and R' we would have (see Figure S.19)

$$|PQ| + |QR| + |PR| = |P'Q| + |QR| + |RP''|$$
$$= |P'P''| \leq |P'Q'| + |Q'R'| + |RP''|$$
$$= |PQ'| + |Q'R'| + |PR'|.$$

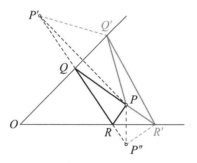

Figure S.19.

5.4 Four, taking into account all possible reflections.

5.5 Orthic triangles degenerate in non-acute triangles. Note that in right triangles, the orthic triangle degenerates to the altitude to the hypotenuse; and in obtuse triangles, some altitudes are exterior to the figure. Thus the minimal solution found for acute triangles does not readily extend to other classes of triangles.

Chapter 6

6.1 Applying Ptolemy's theorem to the cyclic quadrilateral in Figure S.20 yields $|PC| \cdot a = |PA| \cdot a + |PB| \cdot a$.

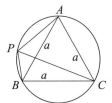

Figure S.20.

6.2 See Figure S.21a.

(a)   (b)

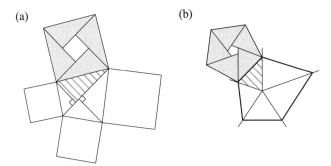

Figure S.21.

6.3 See Figure S.21b.

6.4 Clearly the maximal square should have a vertex on each line. Rotate r
     90° counterclockwise about P to create r', which intersects s at Q (see
     Figure S.22). Now rotate r' clockwise about P (back to r), taking Q to
     Q' on r, with $\angle Q'PQ = 90°$. This determines the solution. Locations
     for Q that produce a greater value of $\overline{PQ}$ yield squares with Q' above
     the line r.

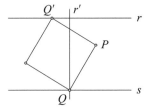

Figure S.22.

6.5 The first inequality follows from Lemma 2.1, while the fourth follows from Challenge 4.3(a). The middle two follow from the AM-GM inequality:

$$ab + bc + ca = a\frac{b+c}{2} + b\frac{c+a}{2} + c\frac{a+b}{2}$$
$$\geq a\sqrt{bc} + b\sqrt{ca} + c\sqrt{ab}, a\sqrt{bc} + b\sqrt{ca} + c\sqrt{ab}$$
$$\geq 3[a\sqrt{bc} \cdot b\sqrt{ca} \cdot c\sqrt{ab}]^{1/3}$$
$$= 3(abc)^{2/3}.$$

6.6 (a) Assume the Weitzenböck inequality holds. Then, using Heron's formula, we have

$$(a^2 + b^2 + c^2)^2 \geq 3 \cdot 16K^2$$
$$= 3(a + b + c)(a + b - c)(a + c - b)(b + c - a)$$
$$= 3\left[2(a^2b^2 + b^2c^2 + c^2a^2) - (a^4 + b^4 + c^4)\right],$$

which is equivalent to the given inequality. Reverse the steps to show that the given inequality implies the Weitzenböck inequality.

(b) See Lemma 2.1.

6.7 Assume $a \leq b \leq c$. If $c^n < a^n + b^n$ for all $n \geq 1$, then $1 < (a/c)^n + (b/c)^n$. Suppose $a \leq b < c$. Then taking limits as n tends to infinity, both $(a/c)^n$ and $(b/c)^n$ tend to zero, yielding the contradiction $1 < 0$. Hence $a \leq b = c$.

Chapter 7

7.1 (a) Apply the Erdős-Mordell inequality in the case when O is the incenter of $\triangle ABC$. Then u, v, and w all equal the inradius r, and $x = r\csc(A/2)$, $y = r\csc(B/2)$, and $z = r\csc(C/2)$. The desired inequality now follows.

(b) Use the facts that $\sin A = a/2R$, $\sin B = b/2R$, $\sin C = c/2R$, and $\sin A + \sin B + \sin C \leq 3\sqrt{3}/2$.

7.2 In Figure S.23 [Bayat et al, 2004], we have $d \leq 1$, so $|a\cos\theta + b\sin\theta|/\sqrt{a^2 + b^2} \leq 1$, from which the desired double inequality follows.

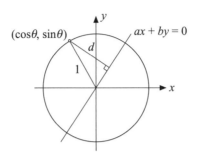

Figure S.23.

The inequalities also follow from the Cauchy-Schwarz inequality applied to the vectors (a, b) and $(\cos\theta, \sin\theta)$.

7.3 Given $a, b > 0$ consider the vectors (a, b) and $(1/2, 1/2)$. Then the Cauchy-Schwarz inequality yields

$$a\frac{1}{2} + b\frac{1}{2} \le \sqrt{a^2 + b^2} \cdot \sqrt{\frac{1}{4} + \frac{1}{4}} = \sqrt{\frac{a^2 + b^2}{2}}.$$

7.4 (a) Expand both sides.

(b) Let $t = 1$ in (a).

(c) Set $t = -1$ in (a) and take square roots to yield

$$\sqrt{a^2 - b^2}\sqrt{x^2 - y^2} = \sqrt{(ax - by)^2 - (ay - bx)^2} \le |ax - by|.$$

7.5 Both solutions are correct as *lower bounds*, but only the second gives the *minimum value* (which is attained at $(a, b, c) = (1/\sqrt{6}, -2/\sqrt{6}, 1/\sqrt{6}$ and similar points).

7.6 Let p_a, p_a, and p_a denote the distances from the circumcenter to the sides a, b, and c. By Carnot's theorem, $p_a + p_b + p_c = R + r$. But (see Figure S.24) $m_a \le R + p_a$, similarly $m_b \le R + p_b$ and $m_c \le R + p_c$. Hence $m_a + m_b + m_c \le 3R + p_a + p_b + p_c = 4R + r$.

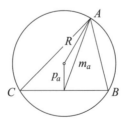

Figure S.24.

7.7 Similarity solves the problem. See Figure S.25.

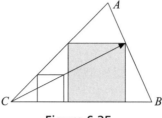

Figure S.25.

7.8 Construct the three altitudes in $\triangle ABC$, and use trigonometry to compute the lengths of the segments of the three sides, as shown in Figure 26a. Then connect the feet of the altitudes to form the orthic triangle, and use the law of cosines to compute the lengths of its sides, as shown in Figure 26b. Thus the three shaded triangles are similar to $\triangle ABC$. If K is the area of $\triangle ABC$, then the areas of the three shaded triangles are $K \cos^2 A$, $K \cos^2 B$, and $K \cos^2 B$. Hence $K \cos^2 A + K \cos^2 B + K \cos^2 C < K$.

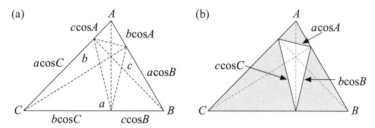

Figure S.26.

Note: One can also (prove and) use the following identity for triangles:

$$\cos^2 A + \cos^2 B + \cos^2 C = 1 - 2 \cos A \cos B \cos C.$$

7.9 Gotman's inequality is equivalent to $\sqrt{2}(a + b)c \le a^2 + b^2 + c^2$. But the Cauchy-Schwarz and AM-GM inequalities yield

$$a \cdot (c\sqrt{2}) + b \cdot (c\sqrt{2}) \le \sqrt{a^2 + b^2} \cdot \sqrt{(c\sqrt{2})^2 + (c\sqrt{2})^2}$$

$$= 2\sqrt{c^2(a^2 + b^2)} \le c^2 + a^2 + b^2.$$

7.10 Let $a, b > 0$ and apply the Cauchy-Schwarz inequality to $(\sqrt{a}, \sqrt{b})$ and $(\sqrt{a}, \sqrt{b})$:

$$2\sqrt{ab} = \sqrt{a}\sqrt{b} + \sqrt{b}\sqrt{a} \le \sqrt{a + b} \cdot \sqrt{a + b} = a + b.$$

7.11 The Cauchy-Schwarz inequality (for two vectors in the plane).

7.12 By the Erdős-Mordell inequality, $3R \ge 2(u + v + w)$.

Chapter 8

8.1 (a) Write $a = \dfrac{a}{a+b}(a+b) + \dfrac{b}{a+b}(0)$ and $b = \dfrac{a}{a+b}(0) + \dfrac{b}{a+b}(a+b)$. Since f is concave, we have

$$f(a) \geq \frac{a}{a+b}f(a+b) + \frac{b}{a+b}f(0),$$

$$f(b) \geq \frac{a}{a+b}f(0) + \frac{b}{a+b}f(a+b),$$

so adding the above inequalities and noting that $f(0) \geq 0$ yields

$$f(a) + f(b) \geq f(a+b) + f(0) \geq f(a+b),$$

i.e. f is subadditive. Part (b) is similar.

8.2 Setting $f(x) = x^{q/p}, 0 < p \leq q$ in Jensen's inequality yields

$$\left[\frac{1}{n}\sum_{i=1}^{n} x_i\right]^{q/p} \leq \frac{1}{n}\sum_{i=1}^{n} x_i^{q/p}.$$

Now let $y_i^p = x_i$:

$$\left[\frac{1}{n}\sum_{i=1}^{n} y_i^p\right]^{q/p} \leq \frac{1}{n}\sum_{i=1}^{n} y_i^q.$$

Taking q-th roots now yields the power mean inequality.

8.3 In Figure 8.13, let $P = (a, f(a))$, $Q = (ta+(1-t)b, f(ta+(1-t)b))$, and $R = (b, f(b))$. If m_{PQ} denotes the slope of the segment PQ (and similarly for m_{PR} and m_{QR}), then

$$m_{PQ} = \frac{f(ta + (1-t)b) - f(a)}{ta + (1-t)b - a} = \frac{f(ta + (1-t)b) - f(a)}{(1-t)(b-a)}$$

$$\leq \frac{tf(a) + (1-t)f(b) - f(a)}{(1-t)(b-a)} = \frac{f(b) - f(a)}{b-a} = m_{PR},$$

and similarly $m_{QR} \geq m_{PR}$. This proves Lemma 8.1.

Now assume f is differentiable. For the points P, Q, R, and S in Figure S.27, we have $m_{PQ} \leq m_{RS}$, or $[f(a) - f(u)]/(a - u) \leq [f(v) - f(b)]/(v - b)$, Taking limits as $u \to a^-$ and $v \to b^+$ yields $f'(a) \leq f'(b)$. This proves Lemma 8.2(i).

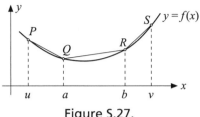

Figure S.27.

To prove Lemma 8.2(ii), assume $x > a$, and apply the mean value theorem to f on $[a, x]$. Then for some c in (a, x),

$$\frac{f(x) - f(a)}{x - a} = f'(c) \geq f'(a)$$

so $f(x) \geq f(a) + f'(a)(x - a)$. When $x < a$ the same inequality is obtained by applying the mean value theorem to f on $[x, a]$.

8.4 (a) Since $1/2\sqrt{n+1} < 1/2\sqrt{n}$, the mediant property yields

$$\frac{1}{2\sqrt{n+1}} < \frac{1+1}{2\sqrt{n+1} + 2\sqrt{n}} < \frac{1}{2\sqrt{n}},$$

but $1/(\sqrt{n+1} + \sqrt{n}) = \sqrt{n+1} - \sqrt{n}$.

(b) Summing the inequalities $\sqrt{n+1} - \sqrt{n} < 1/2\sqrt{n}$ for $n = 1, 2, \ldots, N$ yields $\sqrt{N+1} - 1 < (1/2)\sum_{n=1}^{N} 1/\sqrt{n}$, and hence $\sum_{n=1}^{\infty} 1/\sqrt{n}$ diverges.

8.5 Let $a_n = (1 + 1/n)^n$. Since the natural logarithm is a strictly increasing function, it suffices to show that $\ln a_n < \ln a_{n+1} < 1$ to conclude that $\{a_n\}$ is increasing and bounded above, and thus converges.

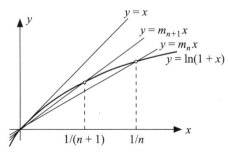

Figure S.28.

Consider the graph of $y = \ln(1 + x)$, as shown in Figure S.28. The slope of the secant line joining (0,0) to $(1/n, \ln(1 + 1/n))$ is $m_n = \ln a_n$. Since the graph of $y = \ln(1 + x)$ is concave (with a tangent line at the origin with slope 1), $m_n < m_{n+1} < 1$, which completes the proof.

8.6 $f(x) = |\sin x|$ is subadditive on $(0, \infty)$ since

$$f(x + y) = |\sin(x + y)| = |\sin x \cos y + \sin y \cos x|$$

$$\leq |\sin x| + |\sin y| = f(x) + f(y).$$

One can draw a figure similar to Figure 8.9 to obtain a geometric interpretation of the subadditivity of f. The function $g(x) = |\cos x|$ is not subadditive on $(0, \infty)$ since $|\cos 2(\pi/2)| = 1 > 0 = 2 |\cos(\pi/2)|$.

8.7 If one superimposes the Lipschitz frame (see Figure 8.8) on the continuity window (see Figure 8.5), one sees that a window of dimensions $[-\delta, \delta] \times [-M\delta, M\delta]$ suffices at every point.

8.8 See Figure S.29 [Brozinski, 1994].

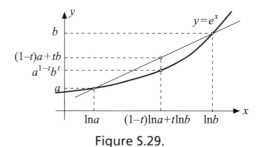

Figure S.29.

Note that the case $t = 1/2$ is the AM-GM inequality.

8.9 See Figure S.30. In the figure, hm, gm, am, rms, and cm denote, respectively, the harmonic, geometric arithmetic, root mean square, and contraharmonic means of a and b.

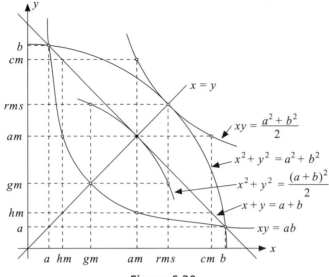

Figure S.30.

8.10 Consider the area under the graph of $y = e^x$ and above the x-axis over the interval $[a,b]$. This area is greater than the area under the tangent line to $y = e^x$ at the point $\big((a + b)/2, e^{(a+b)/2}\big)$ over the same interval.

8.11 Consider $y = e^x$ and the tangent line at its y-intercept.

8.12 $\ln x < x/2$.

8.13 From the graph of f (see Figure S.31) and the computation of f', one finds that $w = (u^p/(u^p + v^p))^{1/q}$ is a critical point of f with $f'(t) > 0$ on $(0, w)$ and $f'(t) < 0$ on $(w, 1)$, so $f(w) = (u^p + v^p)^{1/p}$ is the absolute maximum value of f on $[0,1]$.

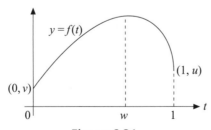

Figure S.31.

Given a_1, a_2, b_1, b_2, set $f(t) = a_1 t + a_2(1 - t^q)^{1/q}$, whose maximum value is $(a_1^p + a_2^p)^{1/p}$. So if we set $t = b_1 / (b_1^q + b_2^q)^{1/q}$, then

$$\frac{a_1 b_1 + a_2 b_2}{(b_1^q + b_2^q)^{1/q}} = f(t) \le f(w) = (a_1^p + a_2^p)^{1/p},$$

which establishes the $n = 2$ case of Hölder's inequality.

8.14 Draw the graph of $F = E - V + 2$ for a fixed V. The proof is analogous to the proof of Theorem 8.1.

8.15 (a) On the graph of $y = \sin x$, consider slopes of secant lines joining $(0,0)$ to $(x, \sin x)$ for x in $(0, \pi)$.

(b) Observe that $n \sin(1/n) = \sin(1/n)/(1/n)$ and use part (a) to show that $\{n \sin(1/n)\}_{n=1}^{\infty}$ is increasing; and consider the tangent line to $y = \sin x$ at the origin to show that $\{n \sin(1/n)\}_{n=1}^{\infty}$ is bounded above.

8.16 Since $f(0) \ge 2f(0)$, we must have $f(0) = 0$. Since $0 \le a \le b$ implies $f(a) \le f(a) + f(b - a) \le f(b)$, f must be nondecreasing. Then, analogous to Figure 8.9, for any $a > 0$ the graph of $y = f(a) + f(x-a)$ on $[a, \infty)$ lies below the graph of $y = f(x)$.

Chapter 9

9.1 See Figure S.32.

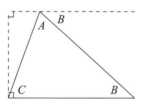

Figure S.32.

9.2 (a) First note that $a \ge b \ge c$ implies $y \ge x \ge z$. Hence $a = x + y \le 2y$. Similarly $a + b = x + 2y + z \le 2x + 2y$. Since $s = x + y + z$, we have $2y \le 2s$ and $2x + 2y \le 2s$.

(b) Observe that (a) is equivalent to $(a, b, c) \prec (a + b - c, a - b + c, -a + b + c)$ and use the fact that $g(x) = \ln x$ is concave.

(c) Use the fact that $g(x) = 1/x$ is convex.

9.3 By symmetry we can assume $a \geq b \geq c$, and draw a figure similar to Figure 2.8, but with heights of the rectangles on the left given by a^k, b^k, c^k rather than a,b,c (note that $a \geq b \geq c$ implies $a^k \geq b^k \geq c^k$) to conclude $ba^k + cb^k + ac^k \leq a^{k+1} + b^{k+1} + c^{k+1}$. We have equality if and only if $a = b = c$.

9.4 Apply the AM-GM inequality twice:

$$\sqrt{(s-b)(s-c)} + \sqrt{bc} \leq \frac{s-b+s-c}{2} + \frac{b+c}{2} = s.$$

9.5 See Figure S.33.

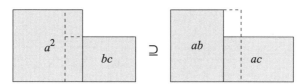

Figure S.33.

9.6 Without loss of generality, assume $a \geq b \geq c$ and $a < b + c$.

(a) Since this inequality is equivalent to $a^2 + b^2 + c^2 \leq (a+b+c)^2/2$, Figure 9.4 will suffice.

(b) This inequality is equivalent to $a^2 + b^2 + c^2 \geq ab + bc + ca$, hence Figure 2.8 suffices. Note that this inequality holds for all $a, b, c \geq 0$.

(c) Redraw Figure 2.17 with $(x_1, x_2, x_3) = (c, b, a)$ and $(y_1, y_2, y_3) = (c^2, b^2, a^2)$.

9.7 We have equality in Bellman's inequality (see Figure 9.2) when $b_i = ka_i$ for some $k > 0$ and $i = 1, 2, L, n$.

9.8 (a) Since $a \geq b \geq c$ implies $A \geq B \geq C$ (Proposition I.19 in the *Elements* of Euclid), Challenge 8.15a gives $\sin C/C \geq \sin B/B \geq \sin A/A$. Using the law of sines ($\sin A/a = \sin B/b = \sin C/c$) now yields the result.

(b) As a consequence of part (a), each summand is nonnegative, hence the sum is as well.

(c) Rearrange the terms in part (b).

(d) Add $A + B + C$ to both sides, and observe that $A + B + C = \pi$ and $a + b + c = 2s$.

9.9 Since $A < \pi/2$, Figure 1.17 yields $\sin A > 2A/\pi$; and from Figure 4.2, $\sin A = a/2R$, so $a/A > 4R/\pi$. Similarly inequalities for b/B and c/C give the result.

9.10 See Figure S.34.

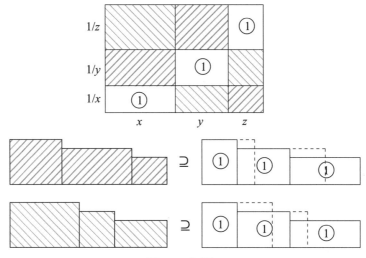

Figure S.34.

Notation and symbols

In this book, the following notation is employed (if not indicated to the contrary in a particular example or application):

$=, \neq$	equal to, not equal to
$>, <$	greater than, less than
$\geq, \leq$	greater than or equal to, less than or equal to
$|a|$	absolute value of the (real) number a
$\|\mathbf{v}\|$	length of the vector $\mathbf{v}$
$\max(a, b)$	largest element of $\{a, b\}$
$\min(a, b)$	smallest element of $\{a, b\}$
$\sqrt{}$	nonnegative square root
$\overline{x}$	arithmetic mean
$\ln x$	natural (or base e) logarithm
AM, GM, HM, RMS	arithmetic mean, geometric mean, harmonic mean, root mean square
P, Q, etc.	points in the plane, space, etc.
PQ	line segment (extended if indicated) joining P and Q
$|PQ|$	length of the line segment PQ
A, B, C	vertices, angles, or measures of the angles of a triangle
$\angle BAC$	angle formed by the segments AB and AC
$ABC, \triangle ABC$	the triangle with vertices A, B, C
a, b, c	lengths of the sides BC, CA, AB ($a = |BC|$, etc.)
L	the perimeter $a + b + c$ of ABC
s	the semiperimeter of ABC, $s = (a + b + c)/2$
K	the area of ABC
I, O	The incenter and circumcenter of ABC
r, R	the inradius and circumradius of ABC

h_a, h_b, h_c	lengths of the altitudes to the sides with lengths a, b, c
m_a, m_b, m_c	lengths of the medians to the sides with lengths a, b, c
w_a, w_b, w_c	lengths of the angle bisectors of angles A, B, C
a, b, c, d	lengths of the sides of a quadrilateral (often named Q)
p, q	lengths of the diagonals of Q
L, K	perimeter and area of Q
$\prec$	majorization order

References

Historical and bibliographical questions are particularly troublesome in a subject like this, which has applications in every part of mathematics but has never been developed systematically.

Hardy, Littlewood and Pólya (1959)

J. Aczél, "A minimum property of the square," in E. F. Beckenbach (ed.), *General Inequalities 2*, Birkhäuser Verlag, Basel (1980), p. 467.

C. Alsina, "A problem," In H. Walter (ed.), *General Inequalities 6*, Birkhäuser Verlag, Basel, (1992), p. 479.

——, "The arithmetic mean-geometric mean inequality for three positive numbers," *Mathematics Magazine*, 73 (2000), p. 97.

——, "Cauchy-Schwarz inequality," *Mathematics Magazine*, 77 (2004a), p. 30.

——, *Contar bien para vivir mejor*, (2nd edition), Editorial Rubes, Barcelona (2004b).

C. Alsina and R. B. Nelsen, *Math Made Visual: Creating Images for Understanding Mathematics*, Mathematical Association of America, Washington, 2006.

——, "On the diagonals of a cyclic quadrilateral," *Forum Geometricorum*, 7 (2007a), pp. 147–149.

——, "A visual proof of the Erdős-Mordell inequality," *Forum Geometricorum*, 7 (2007b), pp. 99–102.

——, "Geometric Proofs of the Weitzenböck and Hadwiger-Finsler Inequalities," *Mathematics Magazine*, 81 (2008), pp. 216–219.

V. I. Arnol'd, *Yesterday and Long Ago*, Springer, Berlin, 2006.

Art of Problem Solving, http://www.artofproblemsolving.com/

A. Avez, "A short proof of a theorem of Erdős and Mordell," *American Mathematical Monthly*, 100 (1993), pp. 60–62.

M. Balinski and H. Young, *Fair Representation: Meeting the Ideal of One Man, One Vote*, 2nd ed., Brookings Institution Press, Washington, 2001.

L. Bankoff, "An elementary proof of the Erdős-Mordell theorem," *American Mathematical Monthly*, 65 (1958), p. 521.

E. J. Barbeau, *Mathematical Fallacies, Flaws and Flimflam*, Mathematical Association of America, Washington, 2000.

M. Bayat, M. Hassani, and H. Teimoori, "Proof without words: Extrema of the function $a \cos t + b \sin t$," *Mathematics Magazine*, 77 (2004), p. 259.

R. Bellman, "A note on inequalities," in E. F. Beckenbarch (ed.), *General Inequalities 1*, Birkhäuser Verlag, Basel (1978), p. 3.

E. Beckenbach and R. Bellman, *An Introduction to Inequalities*, Mathematical Association of America, Washington, 1961.

A. S. Besicovitch, "On Kakeya's Problem and a Similar One". *Math. Z.*, 27 (1928), pp. 312–320.

T. Bonnesen, *Les Problèmes de Isopérimètres et des Isépiphanes*, Paris, Gauthier-Villars, 1929, pp. 59–61.

T. Bonnesen and W. Fenchel, *Theorie der Convexen Körper*, Chelsea Publishing. New York, 1948.

O. Bottema, R. Z. Djordjevič, R. R. Janič, D. S. Mitrionović, P. M. Vasič, *Geometric Inequalities*, Wolters-Nordhoff, Groningen, 1968.

M. K. Brozinski, "Proof without words," *College Mathematics Journal*, 25 (1994), p. 98.

P. S. Bullen, *A Dictionary of Inequalities*, Addison Wesley Longman Ltd., Essex, 1998.

L. Carlitz, "An Inequality Involving the Area of Two Triangles", *American Mathematical Monthly*, 78 (1971), p. 772.

E. Catalan, Note, *J. Ec. Polytechn.*, 24 (1865), pp. 1–71.

R. Courant and H. Robbins, *What is Mathematics?* Oxford University Press, 1941.

H. S. M. Coxeter, "The Lehmus inequality," *Aequationes Mathematicae* 28 (1985), pp. 1–12.

H. S. M. Coxeter and S. L. Greitzer, *Geometry Revisited*, Mathematical Association of America, Washington, 1967.

G. De Santillana, *The Crime of Galileo*, University of Chicago Press, 1955.

N. Dergiades, "An elementary proof of the isoperimetric inequality," *Forum Geometricorum*, 2 (2002), pp. 129–130.

——, "Signed distances and the Erdős-Mordell inequality," *Forum Geometricorum*, 4 (2004), pp. 67–68.

M. de Guzmán, *El rincón de la pizarra. Ensayos de visualización en Análisis Matemático*, Pirámide, Madrid, 1996.

D. Djukić, V. Janković, I. Matić, and N. Petrović, *The IMO Compendium*, Springer, 2006.

H. Dörrie, *100 Great Problems of Elementary Mathematics*, Dover, New York, 1965.

W. Dunham, *Euler, The Master of Us All*, Mathematical Association of America, Washington, 1999.

H. Ehret, "An approach to trigonometric inequalities," *Mathematics Magazine*, 43 (1970), pp. 254–257.

P. Erdős, Problem 3740, *American Mathematical Monthly*, 42 (1935), p. 396.

L. Euler, "Solutio facilis problematum quorundam geometricorum difficillimorum," *Novi Commentarii academiae scientiarum imperialis Petropolitanae* 11 (1767), 103–123. Reprinted in *Opera Omnia*, I 26, pp. 139–157, 1953.

H. Eves, *In Mathematical Circles*, Prindle, Weber & Schmidt, Boston, 1969.

L. Féjes Toth, "On the total lengths of edges of a polyhedron," *Norske, Vid. Selsk. Forch.* 21 (1948), pp. 32–34.

L. Féjes Toth, "Inequalities concerning polygons and polyhedra," *Duke Math. J.*, 54 (1958), pp. 139–146.

Y. Feng, "Jordan's inequality," *Mathematics Magazine*, 69 (1996), p. 126.

R. Ferréol, personal communication (2006).

R. Foote, http://persweb.wabash.edu/facstaff/footer/Planimeter/HowPlanimeters-Work.htm

R. Gelca and T. Andreescu, *Putnam and Beyond*, Springer, 2007.

S. G. Guba, "Zadaca 1797," *Mat. V. Škole* (1977), p. 80.

U. C. Guha, "Arithmetic mean-geometric mean inequality," *Mathematical Gazette*, 51 (1967), pp. 145–146.

G. H. Hardy, J. E. Littlewood and G. Pólya, *Inequalities*, 2nd edition, Cambridge University Press, Cambridge, 1959.

T. L. Heath, *The Works of Archimedes*, Cambridge University Press, 1953.

O. Hölder, *Über einen Mittelwertsatz,* Nachr. Ges. Wiss. Göttingen, 1889.

R. Honsberger, *Mathematical Gems III*, Mathematical Association of America, Washington, 1985.

Human Development Report, 2006: http://hdr.undp.org/en/media/HDR06-complete.pdf (page 395).

W.-D. Jiang, "Proof without words: An algebraic inequality," *Mathematics Magazine*, 80 (2007), p. 344.

D. E. Joyce, *Euclid's Elements.* http://aleph0.clarku.edu/~djoyce/java/elements/elements.html

D. K. Kazarinoff, "A simple proof of the Erdős-Mordell inequality for triangles," *Michigan Math. J.*, 4 (1957), pp. 97–98.

N. D. Kazarinoff, *Geometric Inequalities*, Mathematical Association of America, Washington, 1961.

M. S. Klamkin, "On some geometric inequalities," *Mathematics Teacher*, 60 (1967), pp. 323–328.

Y. Kobayashi, "A geometric inequality," *Mathematical Gazette*, 86 (2002), p. 293.

J. Kocik, "Proof without words: Simpson paradox," *Mathematics Magazine*, 74 (2001), p. 399.

V. Komornik, "A short proof of the Erdős-Mordell theorem," *American Mathematical Monthly*, 104 (1997), pp. 57–60.

S. H. Kung, "Proof without words: The Cauchy-Schwarz inequality," *Mathematics Magazine*, 81 (2008), p. 69.

H. Lee, "Another proof of the Erdős-Mordell theorem," *Forum Geometricorum*, 1 (2001), pp. 7–8.

T. Leise, "As the planimeter's wheel turns: Planimeter proofs for calculus class," *College Mathematics Journal*, 38 (2007), pp. 24–31.

E. Maor, *Trigonometric Delights*, Princeton University Press, Princeton, 1998.

A. Marshall and I. Olkin, *Inequalities: Theory of Majorization and Its Applications*, Academic Press, New York, 1979.

J. V. Martín and Ángel Plaza, "A triangle inequality and its elementary proof," *Math Horizons*, April 2008, p. 30.

H. Minkowski, *Geometrie der Zahlen*, I, Leipzig, 1896.

D. S. Mitrinovic, J. E. Pečarić and Volenec, *Recent Advances in Geometric Inequalities*, Kluwer Academic Pub., Dordrecht, 1989.

D. S. Moore, *The Basic Practice of Statistics*, 3rd ed. W. H. Freeman and Co., New York, 2004.

L. J. Mordell and D. F. Barrow, Solution to Problem 3740, *American Mathematical Monthly*, 44 (1937), pp. 252–254.

R. B. Nelsen, "The harmonic mean-geometric mean-arithmetic mean-root mean square inequality," *Mathematics Magazine*, 60 (1987), p. 158.

——, *Proofs without Words: Exercises in Visual Thinking*, Mathematical Association of America, Washington, 1993. (In Spanish: Proyecto Sur, Granada, 2001).

——, "The Cauchy-Schwarz inequality," *Mathematics Magazine*, 67 (1994a), p. 20.

——, "The sum of a positive number and its reciprocal is at least two (four proofs)," *Mathematics Magazine*, 67 (1994b), p. 374.

——, *Proofs without Words II: More Exercises in Visual Thinking*, Mathematical Association of America, Washington, 2000.

A. M. Nesbitt, "Problem 1514," *Educational Times*, 2 (1903), pp. 37–38.

I. Niven, *Maxima and Minima Without Calculus*, Mathematical Association of America, Washington, 1981.

A. Oppenheim, "The Erdős inequality and other inequalities for a triangle," *American Mathematical Monthly*, 68 (1961), pp. 226–230.

A. Padoa, "Una questione di minimo," *Periodico di Matematiche*, 4 (1925), pp. 80–85.

D. Pedoe, "A two-triangle inequality," *American Mathematical Monthly*, 70 (1963), p. 1012.

——, "Thinking Geometrically", *American Mathematical Monthly*, 77 (1970), pp. 711–721.

——, "Book Review: Recent Advances in Geometric Inequalities," *American Mathematical Monthly*, 98 (1991), pp. 977–980.

H. Poincaré, *Science and Method*, Dover Publications, Inc., New York, 1952.

A. W. Roberts and D. E. Varberg, *Convex Functions*, Academic Press, New York, 1973.

C. Saldaña, *Trastornos del Comportamiento Alimentario*. Terapia de Conducta y Salud, Madrid, 1994.

L. Santaló, "Sobre los sistemas completas de desigualdades entre tres elementos de una figura convexa," *Math. Notae*, XVIII, 1961.

L. Santaló, *Integral Geometry and Geometric Probability*, 2nd ed., Cambridge University Press, Cambridge, 2004.

D. O. Shklarsky, N. N. Chentzov and I. M. Yaglon, *The USSR Olympiad Problem Book*, W. H. Freeman and Co., San Francisco, 1962.

M. Sholander, "On certain minimum problems in the theory of convex curves," *Trans. Amer. Math. Soc.*, 73 (1952), pp. 139–172.

T. A. Sipka, "The law of cosines," *Mathematics Magazine*, 61 (1988), p. 1113.

Stanford Encyclopedia of Philosophy, http://plato.stanford.edu/entries/paradox-simpson/

J. M. Steele, *The Cauchy-Schwarz Master Class*, Mathematical Association of America and Cambridge Univ. Press, Washington-Cambridge, 2004.

G. Steensholt, "Note on an elementary property of triangles," *American Mathematical Monthly*, 68 (1961), pp. 226–230.

D. J. Struik, *A Concise History of Mathematics*, Dover, New York, 1967.

E. Tolsted, "An elementary derivation of the Cauchy, Holder and Minkowski inequalities from Young's inequality," *Mathematics Magazine*, 37 (1964), pp. 2–12.

I. Voicu, "Problem 18666", *Gaz. Mat.*, 86 (1981), p. 112.

E. W. Weisstein, "Cyclic Quadrilateral" From *MathWorld* – A Wolfram Web Resource, http://mathworld.wolfram.com/CyclicQuadrilateral.html

W. H. Young, On classes of summable functions and their Fourier series, *Proc. Royal Soc. (A)*, 87 (1912), pp. 225–229.

Index

About the Authors

Claudi Alsina was born on 30 January 1952 in Barcelona, Spain. He received his BA and PhD in mathematics from the University of Barcelona. His post-doctoral studies were at the University of Massachusetts, Amherst. Claudi, Professor of Mathematics at the Technical University of Catalonia, has developed a wide range of international activities, research papers, publications and hundreds of lectures on mathematics and mathematics education. His latest books include *Associative Functions: Triangular Norms and Copulas* with M.J. Frank and B. Schweizer, WSP, 2006; *Math Made Visual: Creating Images for Understanding Mathematics* (with Roger B. Nelsen), MAA, 2006; *Vitaminas Matemáticas* and *El Club de la Hipotenusa*, Ariel, 2008.

Roger B. Nelsen was born on 20 December 1942 in Chicago, Illinois. He received his BA in mathematics from DePauw University in 1964 and his PhD in mathematics from Duke University in 1969. Roger was elected to Phi Beta Kappa and Sigma Xi. His previous books include *Proofs Without Words: Exercises in Visual Thinking*, MAA 1993; *An Introduction to Copulas*, Springer, 1999 (2nd. ed. 2006); *Proofs Without Words II: More Exercises in Visual Thinking*, MAA, 2000; and *Math Made Visual: Creating Images for Understanding Mathematics* (with Claudi Alsina), MAA, 2006.